Progress in Inflammation Research

Forthcoming titles:

The Resolution of Inflammation, A.G. Rossi, D.A. Sawatzky (Editors), 2007
Angiogenesis in Inflammation: Mechanisms and Clinical Correlates, M.P. Seed, D.A. Walsh (Editors), 2008
New Therapeutic Targets in Rheumatoid Arthritis, P.-P. Tak (Editor), 2008
Inflammatory Cardiomyopathy (DCM) – Pathogenesis and Therapy, H.-P. Schultheiß, M. Noutsias (Editors), 2008
Matrix Metalloproteinases in Tissue Remodelling and Inflammation, V. Lagente, E. Boichot (Editors), 2008

(Already published titles see last page.)

The Immune Synapse as a Novel Target for Therapy

Luis Graca

Editor

Birkhäuser
Basel · Boston · Berlin

Editor

Luis Graca
Unidade de Imunologia Celular
Instituto de Medicina Molecular
Faculdade de Medicina da Universidade de Lisboa
Av. Professor Egas Moniz
1649-208 Lisboa
Portugal

Library of Congress Control Number: 2007934182

Bibliographic information published by Die Deutsche Bibliothek
Die Deutsche Bibliothek lists this publication in the Deutsche Nationalbibliografie;
detailed bibliographic data is available in the internet at http://dnb.ddb.de

ISBN 978-3-7643-8295-7 Birkhäuser Verlag AG, Basel – Boston – Berlin

Printed on acid-free paper produced from chlorine-free pulp. TCF ∞
Cover design: Markus Etterich, Basel
Cover illustration: The immune synapse is often visualized by microscopy using reagents that put in evidence the spatial segregation of molecules involved in antigen-recognition, co-stimulation and adhesion. Visually it resembles a target-like structure with concentric rings. Artist Marta de Menezes created a visual representation of the immune synapse with an image of an eyespot of a *Byciclus anynana* butterfly.
Printed in Germany
ISBN 978-3-7643-8295-7 e-ISBN 978-3-7643-8296-4

9 8 7 6 5 4 3 2 1 www.birkhauser.ch

Contents

List of contributors vii

Preface xi

Emmanuel Donnadieu
The immune synapse and T cell activation: regulation by chemokines 1

Luis Graca
The induction of regulatory T cells by targeting the immune synapse 15

Paul J. Fairchild
Infiltrating the immunological synapse: prospects for the use of altered petide ligands for the treatment of immune pathology 35

Herman Waldmann, Elizabeth Adams and Stephen Cobbold
Targeting CD4 for the induction of dominant tolerance 49

Damien Bresson and Matthias von Herrath
Anti-CD3: from T cell depletion to tolerance induction 57

Yuan Zhai and Jerzy W. Kupiec-Weglinski
Immune modulation by CD40L blockade 71

Francesca Fallarino, Carmine Vacca, Claudia Volpi, Maria T. Pallotta, Stefania Gizzi, Ursula Grohmann and Paolo Puccetti
CTLA-4-immunoglobulin and indoleamine 2,3-dioxygenase in dominant tolerance 87

Mark R. Nicolls and Rasa Tamosiuniene
Adhesion molecules as therapeutic targets 107

Irene Puga and Fernando Macian
E3 ubiquitin ligases and immune tolerance: targeting the immune synapse from within? 129

Bin Li, Xiaomin Song, Arabinda Samanta, Kathryn Bembas, Amy Brown, Geng Zhang, Makoto Katsumata, Yuan Shen, Sandra J. Saouaf and Mark I. Greene
FOXP3 biochemistry will lead to novel drug approaches for vaccines and diseases that lack suppressor T cells 147

Ramireddy Bommireddy and Thomas Doetschman
Transforming growth factor-β: from its effect in T cell activation to a role in dominant tolerance 155

Wang-Fai Ng and John D. Isaacs
From mice to men: the challenges of developing tolerance-inducing biological drugs for the clinic 169

Index 187

List of contributors

Elizabeth Adams, Sir William Dunn School of Pathology, Oxford University, South Parks Road, Oxford OX1 3RE, UK

Kathryn Bembas, Pathology and Laboratory Medicine, University of Pennsylvania, 252 John Morgan, 36th and Hamilton Walk, Philadelphia, PA 19104-6082, USA

Ramireddy Bommireddy, BIO5 Institute and Department of Immunobiology, University of Arizona, PO Box 245217, Tucson, AZ 85724-5217, USA; e-mail: ramiredb@email.arizona.edu

Damien Bresson, La Jolla Institute for Allergy and Immunology, Department of Developmental Immunology 3, 9420 Athena Circle, La Jolla, CA 92037, USA; e-mail: damien@liai.org

Amy Brown, Pathology and Laboratory Medicine, University of Pennsylvania, 252 John Morgan, 36th and Hamilton Walk, Philadelphia, PA 19104-6082, USA

Stephen Cobbold, Sir William Dunn School of Pathology, Oxford University, South Parks Road, Oxford OX1 3RE, UK

Emmanuel Donnadieu, Département de Biologie Cellulaire, Institut Cochin, 22 rue méchain, 75014 Paris, France; e-mail: donnadieu@cochin.inserm.fr

Thomas Doetschman, BIO5 Institute and Department of Cell Biology and Anatomy and Arizona Cancer Center, University of Arizona, PO Box 245217, Tucson, AZ 85724-5217, USA; e-mail: tdoetsch@u.arizona.edu

Paul J. Fairchild, University of Oxford, Sir William Dunn School of Pathology, South Parks Road, Oxford, OX1 3RE, UK; e-mail: paul.fairchild@path.ox.ac.uk

Francesca Fallarino, Dept. of Experimental Medicine, Section of Pharmacology, University of Perugia, Via del Giochetto, 06126 Perugia, Italy; e-mail: fllfnc@tin.it

Stefania Gizzi, Dept. of Experimental Medicine, Section of Pharmacology, University of Perugia, Via del Giochetto, 06126 Perugia, Italy

Mark I. Greene, Pathology and Laboratory Medicine, University of Pennsylvania, 252 John Morgan, 36th and Hamilton Walk, Philadelphia, PA 19104-6082, USA; e-mail: greene@reo.med.upenn.edu

Ursula Grohmann, Dept. of Experimental Medicine, Section of Pharmacology, University of Perugia, Via del Giochetto, 06126 Perugia, Italy

John D. Isaacs, Musculoskeletal Research Group, Institute of Cellular Medicine, University of Newcastle, Newcastle upon Tyne, NE2 4HH, UK; e-mail: j.d.isaacs@newcastle.ac.uk

Makoto Katsumata, Pathology and Laboratory Medicine, University of Pennsylvania, 252 John Morgan, 36th and Hamilton Walk, Philadelphia, PA 19104-6082, USA

Jerzy W. Kupiec-Weglinski, The Dumont-UCLA Transplant Center, Department of Surgery, David Geffen School of Medicine at UCLA, 10833 Le Conte Avenue, Los Angeles, CA 90095, USA

Bin Li, Pathology and Laboratory Medicine, University of Pennsylvania, 252 John Morgan, 36th and Hamilton Walk, Philadelphia, PA 19104-6082, USA

Fernando Macian, Albert Einstein College of Medicine, Department of Pathology, 1300 Morris Park Avenue, Bronx, NY 10461, USA; e-mail: fmacianj@aecom.yu.edu

Wan-Fai Ng, Musculoskeletal Research Group, Institute of Cellular Medicine, University of Newcastle, Newcastle upon Tyne, NE2 4HH, UK

Mark R. Nicolls, Veterans Administration Palo Alto, Health Care System, Medical Service (111), and Division of Pulmonary and Critical Care Medicine, Stanford University, 3801 Miranda Ave., Palo Alto, CA 94304, USA; e-mail: mnicolls@stanford.edu

Maria T. Pallotta, Dept. of Experimental Medicine, Section of Pharmacology, University of Perugia, Via del Giochetto, 06126 Perugia, Italy

Paolo Puccetti, Dept. of Experimental Medicine, Section of Pharmacology, University of Perugia, Via del Giochetto, 06126 Perugia, Italy

Irene Puga, Albert Einstein College of Medicine, Department of Pathology, 1300 Morris Park Avenue, Bronx, NY 10461, USA

Arabinda Samanta, Pathology and Laboratory Medicine, University of Pennsylvania, 252 John Morgan, 36th and Hamilton Walk, Philadelphia, PA 19104-6082, USA

Sandra J. Saouaf, Pathology and Laboratory Medicine, University of Pennsylvania, 252 John Morgan, 36th and Hamilton Walk, Philadelphia, PA 19104-6082, USA

Yuan Shen, Pathology and Laboratory Medicine, University of Pennsylvania, 252 John Morgan, 36th and Hamilton Walk, Philadelphia, PA 19104-6082, USA

Xiaomin Song, Pathology and Laboratory Medicine, University of Pennsylvania, 252 John Morgan, 36th and Hamilton Walk, Philadelphia, PA 19104-6082, USA

Rasa Tamosiuniene, Veterans Administration Palo Alto, Health Care System, Medical Service (111), and Division of Pulmonary and Critical Care Medicine, Stanford University, 3801 Miranda Ave., Palo Alto, CA 94304, USA

Carmine Vacca, Dept. of Experimental Medicine, Section of Pharmacology, University of Perugia, Via del Giochetto, 06126 Perugia, Italy

Claudia Volpi, Dept. of Experimental Medicine, Section of Pharmacology, University of Perugia, Via del Giochetto, 06126 Perugia, Italy

Matthias von Herrath, La Jolla Institute for Allergy and Immunology, Department of Developmental Immunology 3, 9420 Athena Circle, La Jolla, CA 92037, USA

Herman Waldmann, Sir William Dunn School of Pathology, Oxford University, South Parks Road, Oxford OX1 3RE, UK;
e-mail: herman.waldmann@path.ox.ac.uk

Yuan Zhai, The Dumont-UCLA Transplant Center, Department of Surgery, David Geffen School of Medicine at UCLA, 10833 Le Conte Avenue, Los Angeles, CA 90095, USA; e-mail: yzhai@mednet.ucla.edu

Geng Zhang, Pathology and Laboratory Medicine, University of Pennsylvania, 252 John Morgan, 36th and Hamilton Walk, Philadelphia, PA 19104-6082, USA

Preface

It is now accepted that T cell activation by an antigen-presenting cell requires the organization of a supramolecular structure – the immune synapse. This structure, with different types of molecules spatially segregated, is involved in the delivery of quantitative and qualitative signals critical for T cell activation, and therefore in controlling the nature of the immune response. This volume discusses the progress in manipulating components of the immune synapse as a strategy to regulate the immune response in immune pathology, such as transplantation, autoimmunity and allergy.

Donnadieu reviews the current knowledge on the molecular composition and organization of the immune synapse and how the formation of this structure can be modulated by chemokines. It is also known that the immune synapse formation is critical for the activation of naive T cells, as well as their functional polarization. The second chapter discusses the conversion of naive T cells into regulatory T cells (Treg) when components of the immune synapse are manipulated in such a way that the T cells receive suboptimal activation signals.

One way to interfere with the immune synapse's ability to activate T cells is by using altered peptide ligands (APLs). Fairchild reviews the state of the art of immune-modulation with APLs. Several different monoclonal antibodies (mAbs) have been shown to be effective in inducing immune tolerance associated with a dominant role of Treg cells. One of the pioneers of the concept of reprogramming the immune system with non-depleting mAbs is Herman Waldmann. His chapter reviews the use of mAbs targeting CD4 to achieve immune tolerance, giving an historical perspective of the developments in this field. Another T cell co-receptor – the CD3 – has been recently used as a target for tolerogenic mAbs, after many years being mainly used as a target for mAbs leading to T cell depletion, or T cell activation. Bresson and von Herrath review the recent developments in anti-CD3 therapy, with a special emphasis on the treatment of type 1 diabetes.

Tolerance has also been achieved following co-stimulation blockade. Zhai and Kupiec-Weglinski describe how CD40L blockade can lead to long-term transplantation tolerance. Fallarino and colleagues discuss the targeting of CD28 by CTLA-4-Ig and how the tolerance state thus induced appears to relate with changes on tryptophan catabolism triggered by indoleamine 2,3-dioxygenase (IDO) induction. The formation of the immune synapse also requires the participation of cellular adhesion molecules. Nicolls and Tamosiuniene review anti-adhesion therapies – namely with mAbs targeting LFA-1 and ICAM-1 – as immune modulators.

In recent years it became apparent that ubiquitination is an important mechanism in regulating cellular processes. Several E3 ubiquitin ligases, such as Cbl-b, GRAIL and Itch, have been reported as key players in the regulation of immune tolerance. Puga and Macian describe how E3 ubiquitin ligases may modulate the activity of key signaling molecules therefore targeting the immune synapse from within the T cell. Li and colleagues discuss Foxp3 biochemistry, and how drugs interfering with post-translational modification of Foxp3 may control Treg function.

Bommireddy and Doetschman review the role of TGF-β in immune-pathology and in Treg cell function. TGF-β has been reported as a critical cytokine for Treg cell induction in the periphery, as well as being able to function by increasing the threshold necessary for full activation of T cells.

In the final chapter Ng and Isaacs discuss the challenges for translating the knowledge acquired with animal models into tolerance-inducing drugs for the clinic.

I am grateful to all contributors to this volume, who have shared their expertise from basic science to clinical trials. I do hope all readers will be as excited as myself with the promising developments in this field.

Lisboa, August 2007 | Luis Graca

The immune synapse and T cell activation: regulation by chemokines

Emmanuel Donnadieu

Institut Cochin, Université Paris Descartes, CNRS (UMR 8104), Paris, and Inserm, U567, Paris, France

Initial descriptions of the immune synapse

The initiation of an immune response requires that cells communicate with each others to ensure an appropriate protection against pathogens.

T cells do not 'see' free antigens, recognizing them only after processing and presentation by professional antigen-presenting cells (APCs), such as dendritic cells (DCs). This cell-cell contact enables T cells to engage their T cell receptors (TCRs) with cell surface peptide-MHC (pMHC) complexes. Together with signals from costimulation molecules, the antigenic signal leads to T cell activation and differentiation.

The importance of cell communication in the initiation of an immune response was underestimated for a long time. T cell activation was mainly studied using biochemical tools on T cell lines stimulated with antibodies directed against surface molecules. Although these approaches have been very fruitful to dissect the various signaling pathways triggered following TCR stimulation, they did not provide spatio-temporal information at the single-cell level.

However, within the past 15 years, progress in imaging technologies has made it possible to monitor the early consequences of the interaction between T cells and APCs. Initial studies were focused on immediate downstream events triggered in APC-stimulated T cells, including Ca^{2+} responses and shape changes [1, 2].

In 1998, Avi Kupfer and his colleagues [3], using 3-D confocal microscopy, made the striking observation that TCR, other receptors and signaling molecules display a distinct clustering at the T-APC interface. The classical immune synapse (IS) was first characterized in T cell-B cell contacts as a central clustering of small signaling molecules, such as the TCR, CD28 and protein kinase Cθ, that the authors named c-SMAC (for central supramolecular activation cluster), surrounded by a ring of larger adhesion molecules such as the integrin LFA-1, named p-SMAC (for peripheral supramolecular activation cluster).

This report was shortly followed by a study performed with lipids bilayers bearing pMHC complexes and adhesion molecules used as a surrogate for a live APC [4]. This model, although artificial, provided a dynamic view of receptors cluster-

The Immune Synapse as a Novel Target for Therapy, edited by Luis Graca

ing with a good spatio-temporal resolution. It confirmed the partitioning of several surface and intracellular signaling molecules at the c-SMAC and p-SMAC.

Subsequent experiments with green fluorescent protein (GFP)-tagged proteins expressed in T cells have given access to direct dynamic information at the IS and shown some subtle movements of a number of surface and signaling molecules including CD4, CD28, Lck and PI3K [5, 6].

Plasticity of the immune synapse

After these seminal observations performed with B cells as APCs or with planar bilayers, it appeared that the architecture and the composition of the IS were more diverse than initially thought.

Several studies established that the characteristics of the IS are dependent on the nature of the T cells. Hence, immature thymocytes have been shown, in bilayer experiments, to form multiple, small transient clusters of the TCR, but have no central clustering of the TCR or lipid rafts [7]. In addition, murine primary Th1 and Th2 cells differ in the organization of the IS, with Th1 cells, but not Th2 cells, clustering signaling molecules at the T cell/B cell synapse site [8].

Furthermore, the structure of the IS is very dependent on the nature of the APCs. Initial observation of the IS were made with B cells, which in resting state are inefficient APCs. T cells form long-lasting contacts with B cells and as outlined above, c-SMAC and p-SMAC are observed at the interface of both cells.

The situation is strikingly different when one uses DCs, which are the most potent APCs. First, DCs possess a unique ability to trigger multiple signals in the absence of antigen. More than 20 years ago, Nussenzweig and Steinman reported the formation of antigen-independent T-DC conjugates leading to a low level of T cell proliferation [9]. Since this initial description, studies have demonstrated the formation of a functional antigen-independent synapse with the recruitment of surface and signaling molecules at the interface between T cells and DCs [10–13]. Using primary human T cells and autologous DCs, my colleagues and I have shown that DCs are very potent for inducing T lymphocyte motility [14, 15]. Initially round and static, T lymphocytes overlaid onto a monolayer of DCs rapidly adopt a polarized morphology with a leading edge and a uropod. This step is followed by an active motility of T cells that successively interact with multiple DCs. We also found that a fraction of T cells display small and transient Ca^{2+} responses during their migration. Importantly, these antigen-independent signals promote T cell survival [16]. They are also supposed to prime T cells and maintain them in a state of functional alertness ready to detect and rapidly respond to rare expressed cognate antigens [17].

A second feature of the DCs is the absence of a classical c-SMAC and p-SMAC synapse at the contact site with T cells. Of note, in the presence of a cognate antigen, the structure of the synapse formed between murine DCs and naive T cells is multifocal and composed of tight appositions of a few tens of nm in diameter [18]. Interestingly, these numerous tight appositions are reminiscent of the microclusters of TCRs and essential signaling molecules that have been recently described in different models including lipid bilayers [19, 20].

Third, T cells and DCs, even in the presence of a cognate antigen, have a tendency to form dynamic interactions that contrast with the very stable contact between T cells and B cells. By looking at the interaction between T cells and DCs embedded in a collagen matrix, Gunzer et al. [21] showed that T cells only formed short-lived interaction with antigen-loaded DCs. Interestingly, these transient interactions were sufficient to trigger a full T cell activation. However, more recent data indicate that the T-DC contact duration time analyzed in a collagen matrix is very much controlled by the concentration of antigen [22]. Hence, there is a clear positive relation between the amount of antigen and the stability of the interaction between both cell types.

As outlined above, the dynamics and composition of the IS have been studied extensively and a clear picture is emerging from these works. However, there is still considerable debate as to the purpose of the IS. I refer readers to several reviews covering this topic [23, 24].

Studying immune synapse formation and T cell activation *in vivo*

All the studies summarized below were realized in culture systems. Although the *in vitro* condition is convenient and easy to dissect experimentally, the findings obtained in these systems might not always be applicable to the *in vivo* condition. In particular, they did not take into account the complexity of the lymphoid milieu.

Since 2002, the use of two-photon microscopy has enabled several groups to visualize the behavior of T cells and APCs deep within secondary lymphoid organs such as lymph nodes (LNs). These studies indicate that in a non-inflammatory steady-state condition, T cells in LNs migrate rapidly in an apparently random manner [25–27]. This continuous scan may explain the remarkable ability of rare antigen-specific T cells and APCs to find one another to produce an effective immune response. In the presence of a cognate antigen, a stable T-DC contact lasting for several hours is, in most of the published studies, preceded by transient interactions [26, 28, 29]. The rules that control the transition from these brief to stable contacts are not known at present. One can hypothesize that T cells have to reach a certain threshold to make long-lived interactions with DCs. During their transient interactions, T cells would collect and integrate multiple antigen signals but also input from the LN environment. The avidity of the antigen, both its concentration and affinity, is also likely to regulate the duration of the T-DC interaction. At a very high antigen concentration, and similarly to studies performed in culture systems, the T-DC interaction is expected to be immediately stable, although no direct *in vivo* evidence has been published yet. In addition, the maturation of the DCs also influences the nature of the interactions with T cells.

These findings have provided dynamic insights on the initiation of the immune response in LNs. However, they were essentially focused on the motile behavior of T cells in the absence and presence of antigen and, until now, only a few studies have analyzed the assembly and structure of the IS *in vivo*. Stoll et al. [30] used a conventional confocal approach to analyze the behavior of antigen-specific T cells expressing CD43 tagged to GFP. CD43 is a large glycoprotein found, in culture systems, to be excluded from the IS. It is striking to note that the same pattern of exclusion was also observed in T cells during their interactions with DCs within LNs. More recently, Barcia et al. [31] analyzed the interaction between effector CD8 T cells and virally infected brain astrocytes. They confirmed a molecular partitioning composed of both c and p-SMAC at the interface between T cells and target cells. However, these studies realized on fixed tissues only provided static snapshots of highly dynamic processes.

To characterize T cell responses measured in a quasi-natural environment, my colleagues and I have developed a new experimental system of thin slices of LNs [32], adapted from the recently described method for thymic slices [33]. In this system, fluorescently labeled T cells overlaid onto a LN slice are rapidly recruited into the tissue and display a vigorous motility. This system enabled us to monitor other responses than T cell motility. Notably, the fast recruitment of T cell into the slice allows us to use the fluorescent Ca^{2+} dye fura-2, which has a tendency to leak out of the cells over time, thus limiting its use after adoptive transfer experiments. Our data indicate that a substantial proportion of T cells migrating within the slice display small and transient Ca^{2+} responses similar to those observed *in vitro* when T cells interact with DCs.

Overall, these studies performed in a near physiological T cell environment established that during their displacements within an LN, T cells are continuously exposed to diverse extracellular inputs that will have an important impact on the IS and will fine-tune T cell activation thresholds.

Chemokines: more than attracting molecules

Chemokines are important in establishing the distribution of lymphocyte subpopulations in primary and secondary lymphoid tissues and in the recruitment of leukocytes to sites of inflammation. For example, the recruitment of naive T cells in LNs is controlled by CCR7 ligands, CCL19 and CCL21 [34].

However, several findings indicate that chemokines possess other functions than their chemoattractant activities. These molecules have been shown to positively regulate the synapse between T cells and DCs. First evidence was obtained studying the consequences of the interaction between human primary T cells and autologous DCs [14, 15]. In this system, we found that DC-induced T cell responses observed in the absence of antigen, i.e., motility and Ca^{2+} increases, were dependent on chemokines secreted by DCs. Additional experiments indicated that immature DCs stimulate primary CD4 memory T cells through the secretion of CCR4 ligands, CCL17 and CCL22 [14]. During their maturation, DCs secrete a whole array of chemokines including CCL19, which polarize naive T cells, and trigger a motility signal that enables these cells to find rare expressed antigens [15].

A similar situation has been reported with murine T cells and DCs but with slight differences [35]. In this case, CCL21 but not CCL19 was the important chemokine expressed at the surface of DCs purified from LNs. Friedman et al. [35] reported a two-step contact mechanism where T cells overlaid onto a monolayer of DCs rapidly get polarized and adhere to the APC through their uropods. Then, T cells use their leading edge to efficiently probe their surroundings and contact other DCs.

Recently, the importance of chemokines in T cell scanning and motility was confirmed in more physiological settings. Three groups including our own have shown that CCR7 ligands, CCL19 and CCL21, play important roles in driving T cell motility within LNs slices or intact organs [32, 36, 37]. These studies took advantage of a mouse strain (PLT) that is deficient in CCL19 and CCL21 as well as CCR7-deficient mice. In the absence of CCR7 or its ligands, T cell migration was partially inhibited by 50%. Of note, the residual T cell motility was sensitive to pertussin toxin known to inactivate the Gαi proteins used by most chemokine receptors to transmit their signals. Thus, intranodal T cell migration is regulated by CCR7 ligands and other unknown factors that await future identification.

Another important issue concerns the nature of the cells producing these lymphoid chemokines. Previous studies by Jason Cyster and his colleagues [38] revealed that stromal cells of the T zone, called fibroblastic reticular cells (FRCs), produce large quantities of chemokines including CCL19 and CCL21. In addition, a recent study performed in intact LN has shown that T cells did not migrate totally randomly within the node but were in fact guided in their displacements by FRCs [39]. Thus, FRCs can be considered as railways providing both a support and a chemokinetic trigger to T cells. According to the current model, the stromal networks guide T cells towards DCs allowing lymphocytes to interact with multiple DCs also positioned along FRCs [40] (Fig. 1).

Does this mean that DCs have only a passive role waiting for fast traveling T cells on FRCs? I believe that the situation is more complex and that DCs could also contribute to the T cell exploratory behavior. As previously shown *in vitro*, chemokines expressed by DCs might promote a motility signal that would help T cells to efficiently scan the surface of the APC. On the other hand, chemokines secreted by DCs could act in a more classical manner by attracting T cells. Such a chemotactic guidance cue has been shown to occur after the initiation of an immune response. Indeed, antigen-dependent conjugates of CD4 T cells and DCs have the potential to

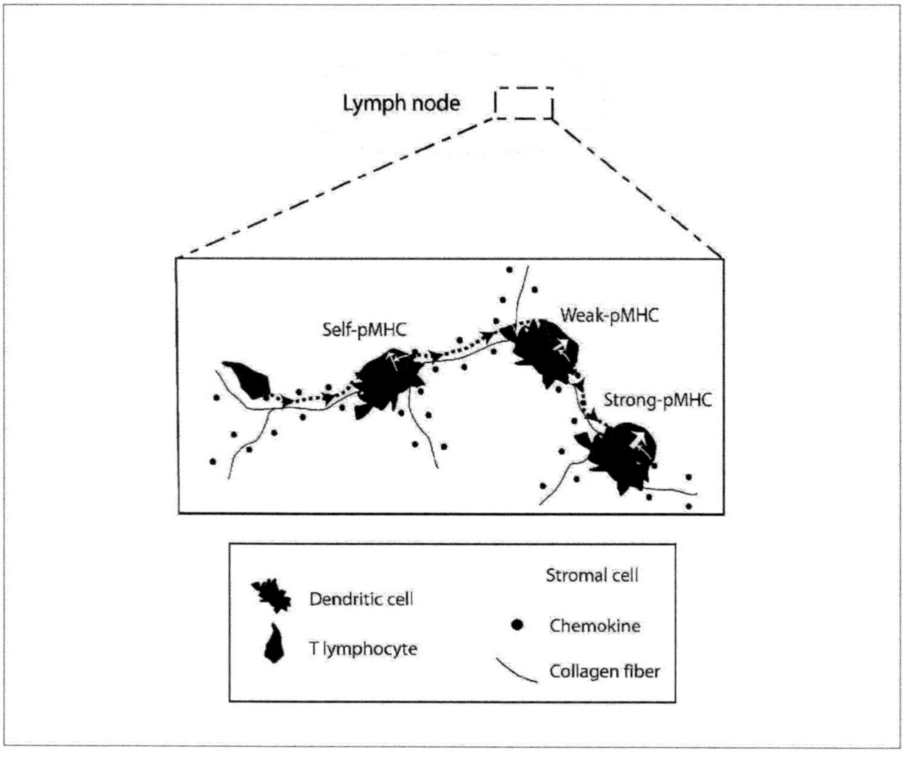

Figure 1

Model of T cell migration and immune synapse (IS) formation in the lymph node (LN). Upon entering a LN, T lymphocytes interact with fibroblastic reticular cells (FRCs) and move along these cells. Stromal cells provide both an adhesion substrate and chemokines immobilized at their surfaces enabling and restricting T cell migration. Along their paths, T cells make short-lived interactions with dendritic cells (DCs) looking for specific MHC-peptide (pMHC) complexes. Even if no cognate antigen is detected, DCs trigger several low intensity signals to T cells. These signals from self antigens, adhesion and soluble molecules including chemokines maintain T cells in a state of functional alertness that favor antigen detection and responsiveness. The situation is different when T cells recognize specific pMHC and the duration of T-DC interaction is most likely dependent on the overall avidity of the antigen being presented by DCs. At low cognate antigen dose, T cells engage first in a series of short interactions with multiple DCs then make stable interaction with the antigen-presenting cells (APCs). In contrast, at high antigen dose, immediate stable T-DC conjugates occur. During their interactions with DCs, T cells integrate multiple extracellular inputs from the TCR, costimulation and adhesion molecules but also soluble molecules including chemokines and cytokines leading to full activation.

attract CD8 T cells through the secretion of CCL3 and CCL4 inflammatory chemokines [41]. Whether DCs also generate hot spots of attractiveness in non-inflammatory steady-state conditions is an open question.

Independent of their chemoattractant and guiding activities, chemokines can also modulate the structure and functioning of the IS and consequently the outcome of an immune response. Over the last few years, several reports have analyzed the role of chemokines during T cell-APC interaction. Looking at the IS formed between Jurkat T cells and B cell lines, Viola and co-workers [42] found that CCR5 and CXCR4 as well as their ligands were recruited to the IS, resulting in stable T-APC conjugates and enhancement of T cell proliferation. An increase in antigen-dependent T cell responses by chemokines was also reported recently in murine models [35]. Therefore, it is becoming clear that IS formation and functioning is not only controlled by adhesion and costimulation molecules but also by chemokines. Reminiscent of the neuronal synapses where neurons use neurotransmitters for their communication, chemokines could be considered as immunotransmitters [43].

Several questions arise from these studies. First, the mechanisms involved in these effects are poorly defined. It has long been known that chemokines are able to costimulate T cells *in vitro*. Hence, multiple chemokines including CCL2, CCL3, CCL4 and CXCL12 are capable of increasing T cell responses (e.g., proliferation, secretion of cytokines) induced by anti-CD3 antibodies [44, 45]. Indeed, chemokines trigger a signaling cascade that together with that of TCR could lead to full T cell activation.

Chemokines are also well known to induce a rapid increase in the avidity of the integrin LFA-1 that may stabilize T cell-APC conjugates and consequently enhance T cell responses. In addition, chemokines trigger T cell polarization, which is characterized by accumulation of adhesion, costimulation, and signaling molecules at the leading edge, with exclusion of molecules that impair cell-cell contacts such as CD43. Hence, previous observations have shown that T cells were much more sensitive to APC contact made at the leading edge of the cell than with contact made at the tail [46]. A new publication has extended this further and confirmed the notion that chemokines lower the activation threshold of polarized T cells [35]. Whether chemokines also enhance T cell responses to cognate antigens *in vivo* is not known and await further investigations. It would add to the already multiple functions of chemokines in the regulation of an immune response.

Altogether, these studies suggest that chemokines positively regulate both antigen detection as well as IS formation. However, chemokines also display other effects on T cells that need to be mentioned. First, under certain chemokine gradients the interaction between T cells and MHC-peptide complexes have been shown to be of short duration. This premature separation results in a down-modulation of T cell proliferation [47]. As these results have been obtained *in vitro* with transwell chambers, it will be important to know if comparable chemokine gradients also occur in the LN and affect the IS formation.

Second, a few chemokines including CXCL12 can provoke an active movement of leukocytes away from this stimulus [48, 49]. These chemorepulsive effects, which have been identified at high concentrations of CXCL12, are supposed to play a key role in T cell emigration from the thymus [50]. Again, it is interesting to draw a parallel with these studies and those performed in the nervous system where molecules established to have repellent functions, such as members of the ephrin and semaphorin families, have been previously described and extensively studied.

Whether factors other than CXCL12 harbor the same repulsive effect on T cells is not known. Furthermore, the physiological importance of such repulsion in the T cell behavior studied within LNs and others tissues remains an open question.

Cytokines

T cells are also influenced by cytokine signals from the environment. Cytokines are recognized as key players in T cell survival, proliferation and differentiation. Despite this, very few things are known on the role that these growth factors might play in the formation of the IS.

Within an LN, FRCs and DCs secrete multiple cytokines that might regulate early T cell responses. For example, cytokines such as IL-2, IL-6 and IL-15 have been shown to trigger a signaling cascade leading to T cell polarization and migration [51–53]. However, most of these studies have been performed on activated or memory T cells and it is not known if cytokines such as IL-7 that act on naive T cell possess the potential to polarize T lymphocytes. Once an IS is formed, cytokines secreted by newly activated T cells could affect both partners in an autocrine and paracrine manner. Indeed, seminal studies from Kupfer and co-workers [54], recently revisited by Huse et al. [55], performed with B cells as APCs and activated helper T cells, have provided evidence that several cytokines including IL-2 were secreted into the synaptic cleft, whereas others were released multidirectionally. In addition, the presence or not of certain cytokine receptors at the IS have been suggested to directly influence the differentiation of helper T cells [56]. Clearly, more work needs to be done to address the role of cytokines in IS assembly and function.

Regulatory T cells

Regulatory T cells (Treg cells) are another component of the immune system that plays a prominent role in down-modulating T cell responses. Despite extensive research on these cells, the molecular mechanisms by which Treg cells inhibit immune response remain largely elusive. Initial studies have focused on soluble factors and in some cases, Treg cells seem to suppress *via* IL-10 [57]. Recent experimental data obtained *in vitro* but also *in vivo* shed new light and indicate that Treg cells inhibit earliest T cell responses and T-DC contact formation. In a study performed in culture systems, Treg cells have been shown to affect autoreactive T cells by decreasing the recruitment of the PKCθ to the synapse formed with APCs [58]. Since this enzyme is directly coupled to NF-κB activation, this could be one of the targets of Treg cells responsible for decreasing T cell activation.

An early effect of Treg cells on the synapse formed between T cells and DCs have been confirmed recently by two-photon microscopy of intact LNs. These studies demonstrate that Treg cells prevent stable contact formation between DCs and autoreactive T cells leading to T cell priming [59, 60]. Interestingly, Treg cells do not interact with autoreactive T cells but formed long lasting conjugates with DCs. These findings support the notion that part of Treg cell suppression is mediated through DCs.

Future directions

Nearly 10 years after the initial description of the immune synapse, many issues remain unresolved. As outlined above, the dynamics and composition of the IS in culture systems have been studied extensively. On the other hand, two-photon microscopy has provided important insights on the motile behavior of immune cells analyzed in their natural environment before and after antigen recognition. This novel and powerful method coupled to genetic approaches to express fluorescent proteins will allow the visualization, in real-time, of the IS formation *in vivo*. Is a recruitment of molecules observed during the transient T-DC interaction that occurs rapidly after antigen detection? Are c- and p-SMAC structures also observed in these conditions?

In addition, the measurement of functional read-outs at the single-cell level will be particularly relevant to address the question of how long a T cell needs to contact an APC to be fully activated in an LN. The use of fluorescent probes such as the Ca^{2+} dye fura-2, as well as gene expression studies with fluorescent molecules coupled to the IL-2 promoter should significantly advance our understanding in this field of research.

All of these issues will benefit from progress in instrumentation/detection as, until now, two-photon microscopy detects only high-intensity signals. So far, the T cell behavior has been analyzed mostly exclusively in LNs and only a few studies have been extended to non-lymphoid tissues. The combination of different physiological systems such as intact organs, but also slices of tissues will enable the measurement of T cell responses and IS formation also in effector sites. This should yield progress in various areas such as autoimmunity and cancer.

References

1 Donnadieu E, Bismuth G, Trautmann A (1994) Antigen recognition by helper T cells elicits a sequence of distinct changes of their shape and intracellular calcium. *Curr Biol* 4: 584–595

2 Negulescu PA, Krasieva TB, Khan A, Kerschbaum HH, Cahalan MD (1996) Polarity of T cell shape, motility, and sensitivity to antigen. *Immunity* 4: 421–430

3 Monks CR, Freiberg BA, Kupfer H, Sciaky N, Kupfer A (1998) Three-dimensional segregation of supramolecular activation clusters in T cells. *Nature* 395: 82–86

4 Grakoui A, Bromley SK, Sumen C, Davis MM, Shaw AS, Allen PM, Dustin ML (1999) The immunological synapse: a molecular machine controlling T cell activation. *Science* 285: 221–227

5 Harriague J, Bismuth G (2002) Imaging antigen-induced PI3K activation in T cells. *Nat Immunol* 3: 1090–1096

6 Zal T, Zal MA, Gascoigne NR (2002) Inhibition of T cell receptor-coreceptor interactions by antagonist ligands visualized by live FRET imaging of the T-hybridoma immunological synapse. *Immunity* 16: 521–534

7 Hailman E, Burack WR, Shaw AS, Dustin ML, Allen PM (2002) Immature $CD4^+CD8^+$ thymocytes form a multifocal immunological synapse with sustained tyrosine phosphorylation. *Immunity* 16: 839–848

8 Balamuth F, Leitenberg D, Unternaehrer J, Mellman I, Bottomly K (2001) Distinct patterns of membrane microdomain partitioning in Th1 and th2 cells. *Immunity* 15: 729–738

9 Nussenzweig MC, Steinman RM (1980) Contribution of dendritic cells to stimulation of the murine syngeneic mixed leukocyte reaction. *J Exp Med* 151: 1196–1212

10 Delon J, Bercovici N, Raposo G, Liblau R, Trautmann A (1998) Antigen-dependent and -independent Ca^{2+} responses triggered in T cells by dendritic cells compared with B cells. *J Exp Med* 188: 1473–1484

11 Montes M, McIlroy D, Hosmalin A, Trautmann A (1999) Calcium responses elicited in human T cells and dendritic cells by cell-cell interaction and soluble ligands [published erratum appears in *Int Immunol* (1999) 11: 1275–1276]. *Int Immunol* 11: 561–568

12 Revy P, Sospedra M, Barbour B, Trautmann A (2001). Functional antigen-independent synapses formed between T cells and dendritic cells. *Nat Immunol* 2: 925–931

13 Kondo T, Cortese I, Markovic-Plese S, Wandinger KP, Carter C, Brown M, Leitman S, Martin R (2001) Dendritic cells signal T cells in the absence of exogenous antigen. *Nat Immunol* 2: 932–938

14 Real E, Kaiser A, Raposo G, Amara A, Nardin A, Trautmann A, Donnadieu E (2004) Immature dendritic cells (DCs) use chemokines and intercellular adhesion molecule (ICAM)-1, but not DC-specific ICAM-3-grabbing nonintegrin, to stimulate $CD4^+$ T cells in the absence of exogenous antigen. *J Immunol* 173: 50–60

15 Kaiser A, Donnadieu E, Abastado JP, Trautmann A, Nardin A (2005) CC chemokine

ligand 19 secreted by mature dendritic cells increases naive T cell scanning behavior and their response to rare cognate antigen. *J Immunol* 175: 2349–2356

16 Feuillet V, Lucas B, Di Santo JP, Bismuth G, Trautmann A (2005) Multiple survival signals are delivered by dendritic cells to naive CD4+ T cells. *Eur J Immunol* 35: 2563–2572

17 Stefanova I, Dorfman JR, Germain RN (2002) Self-recognition promotes the foreign antigen sensitivity of naive T lymphocytes. *Nature* 420: 429–434

18 Brossard C, Feuillet V, Schmitt A, Randriamampita C, Romao M, Raposo G, Trautmann A (2005) Multifocal structure of the T cell–dendritic cell synapse. *Eur J Immunol* 35: 1741–1753

19 Yokosuka T, Sakata-Sogawa K, Kobayashi W, Hiroshima M, Hashimoto-Tane A, Tokunaga M, Dustin ML, Saito T (2005) Newly generated T cell receptor microclusters initiate and sustain T cell activation by recruitment of Zap70 and SLP-76. *Nat Immunol* 6: 1253–1262

20 Varma R, Campi G, Yokosuka T, Saito T, Dustin ML (2006) T cell receptor-proximal signals are sustained in peripheral microclusters and terminated in the central supramolecular activation cluster. *Immunity* 25: 117–127

21 Gunzer M, Schafer A, Borgmann S, Grabbe S, Zanker KS, Brocker EB, Kampgen E, Friedl P (2000) Antigen presentation in extracellular matrix: interactions of T cells with dendritic cells are dynamic, short lived, and sequential. *Immunity* 13: 323–332

22 Rothoeft T, Balkow S, Krummen M, Beissert S, Varga G, Loser K, Oberbanscheidt P, van den Boom F, Grabbe S (2006) Structure and duration of contact between dendritic cells and T cells are controlled by T cell activation state. *Eur J Immunol* 36: 3105–3117

23 Davis DM, Dustin ML (2004) What is the importance of the immunological synapse? *Trends Immunol* 25: 323–327

24 Trautmann A, Valitutti S (2003) The diversity of immunological synapses. *Curr Opin Immunol* 15: 249–254

25 Miller MJ, Wei SH, Parker I, Cahalan MD (2002) Two-photon imaging of lymphocyte motility and antigen response in intact lymph node. *Science* 296: 1869–1873

26 Mempel TR, Henrickson SE, Von Andrian UH (2004) T-cell priming by dendritic cells in lymph nodes occurs in three distinct phases. *Nature* 427: 154–159

27 Bousso P, Robey E (2003) Dynamics of CD8+ T cell priming by dendritic cells in intact lymph nodes. *Nat Immunol* 4: 579–585

28 Miller MJ, Safrina O, Parker I, Cahalan MD (2004) Imaging the single cell dynamics of CD4+ T cell activation by dendritic cells in lymph nodes. *J Exp Med* 200: 847–856

29 Shakhar G, Lindquist RL, Skokos D, Dudziak D, Huang JH, Nussenzweig MC, Dustin ML (2005) Stable T cell-dendritic cell interactions precede the development of both tolerance and immunity *in vivo*. *Nat Immunol* 6: 707–714

30 Stoll S, Delon J, Brotz TM, Germain RN (2002) Dynamic imaging of T cell-dendritic cell interactions in lymph nodes. *Science* 296: 1873–1876

31 Barcia C, Thomas CE, Curtin JF, King GD, Wawrowsky K, Candolfi M, Xiong WD, Liu C, Kroeger K, Boyer O et al (2006) *In vivo* mature immunological synapses forming

SMACs mediate clearance of virally infected astrocytes from the brain. *J Exp Med* 203: 2095–2107
32 Asperti-Boursin F, Real E, Bismuth G, Trautmann A, Donnadieu E (2007) CCR7 ligands control basal T cell motility within lymph node slices in a phosphoinositide 3-kinase-independent manner. *J Exp Med* 204: 1167–1179
33 Bhakta NR, Oh DY, Lewis RS (2005) Calcium oscillations regulate thymocyte motility during positive selection in the three-dimensional thymic environment. *Nat Immunol* 6: 143–151
34 Cyster JG (2005) Chemokines, sphingosine-1-phosphate, and cell migration in secondary lymphoid organs. *Annu Rev Immunol* 23: 127–159
35 Friedman RS, Jacobelli J, Krummel MF (2006) Surface-bound chemokines capture and prime T cells for synapse formation. *Nat Immunol* 7: 1101–1108
36 Okada T, Cyster JG (2007) CC chemokine receptor 7 contributes to Gi-dependent T cell motility in the lymph node. *J Immunol* 178: 2973–2978
37 Worbs T, Mempel TR, Bolter J, von Andrian UH, Forster R (2007) CCR7 ligands stimulate the intranodal motility of T lymphocytes *in vivo*. *J Exp Med* 204: 489–495
38 Luther SA, Tang HL, Hyman PL, Farr AG, Cyster JG (2000) Coexpression of the chemokines ELC and SLC by T zone stromal cells and deletion of the ELC gene in the plt/plt mouse. *Proc Natl Acad Sci USA* 97: 12694–12699
39 Bajenoff M, Egen JG, Koo LY, Laugier JP, Brau F, Glaichenhaus N, Germain RN (2006) Stromal cell networks regulate lymphocyte entry, migration, and territoriality in lymph nodes. *Immunity* 25: 989–1001
40 Mempel TR, Junt T, von Andrian UH (2006) Rulers over randomness: stroma cells guide lymphocyte migration in lymph nodes. *Immunity* 25: 867–869
41 Castellino F, Huang AY, Altan-Bonnet G, Stoll S, Scheinecker C, Germain RN (2006) Chemokines enhance immunity by guiding naive $CD8^+$ T cells to sites of $CD4^+$ T cell-dendritic cell interaction. *Nature* 440: 890–895
42 Molon B, Gri G, Bettella M, Gomez-Mouton C, Lanzavecchia A, Martinez AC, Manes S, Viola A (2005) T cell costimulation by chemokine receptors. *Nat Immunol* 6: 465–471
43 Trautmann A (2005) Chemokines as immunotransmitters? *Nat Immunol* 6: 427–428
44 Nanki T, Lipsky PE (2000) Cutting edge: Stromal cell-derived factor-1 is a costimulator for $CD4^+$ T cell activation. *J Immunol* 164: 5010–5014
45 Stachowiak AN, Wang Y, Huang YC, Irvine DJ (2006) Homeostatic lymphoid chemokines synergize with adhesion ligands to trigger T and B lymphocyte chemokinesis. *J Immunol* 177: 2340–2348
46 Wei X, Tromberg BJ, Cahalan MD (1999) Mapping the sensitivity of T cells with an optical trap: polarity and minimal number of receptors for Ca(2+) signaling. *Proc Natl Acad Sci USA* 96: 8471–8476
47 Bromley SK, Peterson DA, Gunn MD, Dustin ML (2000) Cutting edge: Hierarchy of chemokine receptor and TCR signals regulating T cell migration and proliferation. *J Immunol* 165: 15–19

48 Poznansky MC, Olszak IT, Foxall R, Evans RH, Luster AD, Scadden DT (2000) Active movement of T cells away from a chemokine. *Nat Med* 6: 543–548

49 Poznansky MC, Olszak IT, Evans RH, Wang Z, Foxall RB, Olson DP, Weibrecht K, Luster AD, Scadden DT (2002) Thymocyte emigration is mediated by active movement away from stroma-derived factors. *J Clin Invest* 109: 1101–1110

50 Vianello F, Kraft P, Mok YT, Hart WK, White N, Poznansky MC (2005) A CXCR4-dependent chemorepellent signal contributes to the emigration of mature single-positive CD4 cells from the fetal thymus. *J Immunol* 175: 5115–5125

51 Nieto M, del Pozo MA, Sanchez-Madrid F (1996) Interleukin-15 induces adhesion receptor redistribution in T lymphocytes. *Eur J Immunol* 26: 1302–1307

52 Arrieumerlou C, Donnadieu E, Brennan P, Keryer G, Bismuth G, Cantrell D, Trautmann A (1998) Involvement of phosphoinositide 3-kinase and Rac in membrane ruffling induced by IL-2 in T cells. *Eur J Immunol* 28: 1877–1885

53 Weissenbach M, Clahsen T, Weber C, Spitzer D, Wirth D, Vestweber D, Heinrich P C, Schaper F (2004) Interleukin-6 is a direct mediator of T cell migration. *Eur J Immunol* 34: 2895–2906

54 Kupfer A, Mosmann TR, Kupfer H (1991) Polarized expression of cytokines in cell conjugates of helper T cells and splenic B cells. *Proc Natl Acad Sci USA* 88: 775–779

55 Huse M, Lillemeier BF, Kuhns MS, Chen DS, Davis MM (2006) T cells use two directionally distinct pathways for cytokine secretion. *Nat Immunol* 7: 247–255

56 Maldonado RA, Irvine DJ, Schreiber R, Gimcher LH (2004) A role for the immunological synapse in lineage commitment of CD4 lymphocytes. *Nature* 431: 527–532

57 Asseman C, Mauze S, Leach MW, Coffman RL, Powrie F (1999) An essential role for interleukin 10 in the function of regulatory T cells that inhibit intestinal inflammation. *J Exp Med* 190: 995–1004

58 Sumoza-Toledo A, Eaton AD, Sarukhan A (2006) Regulatory T cells inhibit protein kinase C theta recruitment to the immune synapse of naive T cells with the same antigen specificity. *J Immunol* 176: 5779–5787

59 Tadokoro CE, Shakhar G, Shen S, Ding Y, Lino AC, Maraver A, Lafaille JJ, Dustin ML (2006) Regulatory T cells inhibit stable contacts between $CD4^+$ T cells and dendritic cells *in vivo*. *J Exp Med* 203: 505–511

60 Tang Q, Adams JY, Tooley AJ, Bi M, Fife BT, Serra P, Santamaria P, Locksley RM, Krummel MF, Bluestone JA (2006) Visualizing regulatory T cell control of autoimmune responses in nonobese diabetic mice. *Nat Immunol* 7: 83–92

The induction of regulatory T cells by targeting the immune synapse

Luis Graca

Instituto de Medicina Molecular, Faculdade de Medicina, Universidade de Lisboa, and Instituto Gulbenkian de Ciência, Oeiras, Portugal

Introduction

The control of deleterious immune responses causing diseases, such as allergy, autoimmunity and transplant rejection, has been one of the main objectives of immunologists. In recent years, regulatory T (Treg) cells have reached center-stage in immunology research, as they were shown to be effective in preventing deleterious immune responses and maintaining a healthy state of immunological non-responsiveness, also known as immune tolerance [1]. It is now accepted that a natural population of Treg cells is generated in the thymus and plays a major role in preventing autoimmunity [2]. Several research groups have been developing strategies to isolate and expand rare self-reactive Treg cell clones as a therapeutic strategy for autoimmunity [3, 4].

As an alternative to the expansion of rare Treg cells, several research groups have established that naive non-regulatory precursors in the periphery can be converted to Treg cells (i.e. independently from the thymus) [5–7]. Such conversion has been observed both *in vitro* and *in vivo*.

In addition, several therapeutic strategies have been developed to induce a state of dominant tolerance in autoimmunity or transplantation by interfering with molecules involved in the formation of the immune synapse and thus in T cell activation, as detailed in the other chapters of this volume. It should be noted that the state of dominant tolerance induced by these reagents appears to require the presence of Treg cells and, in several cases, a causal relationship between tolerogenic treatment and peripheral conversion of Treg cells has been established (discussed below).

The induction of dominant tolerance

It has been documented that several therapeutic strategies are effective in inducing a state of dominant tolerance in experimental animals [8]. It has been shown that several monoclonal antibodies (mAbs) targeting T cell surface molecules, such as

The Immune Synapse as a Novel Target for Therapy, edited by Luis Graca

Table 1 - Monoclonal antibodies in transplantation.

Antibody	Transplantation	Ref.
CD2	Delayed rejection of islet allografts and xenografts	[85]
CD3	Tolerance to heart allografts	[86]
CD3+ CD2	Long-term acceptance to heart grafts	[87]
CD4	Tolerance to antigens	[9, 88]
CD4+CD8	Tolerance to minor mismatched skin, MHC-mismatched and xenogeneic heart grafts	[9, 89]
CD4+CD8+CD154	Tolerance to MHC-mismatched skin	[25]
CTLA4-Ig	Xenogeneic pancreatic islets	[90]
CD45RB	Long-term survival of MHC-mismatched islets and kidney transplants	[91, 92]
CD45RB+ CD154	Long-term survival of MHC-mismatched skin	[93]
CD134L (OX40L)	Long-term survival of minor mismatched heart allografts	[94]
CD134L+ CTLA4-Ig	Long-term acceptance of MHC-mismatched heart grafts	[95]
CD154	MHC-mismatched pancreatic islets	[96]
CD154+CD8dep	Tolerance to minor mismatched skin	[10, 97]
CD154+CTLA4-Ig	Long-term acceptance MHC-mismatched skin	[98, 99]
LFA-1	Tolerance to soluble antigen	[100]
LFA-1+ICAM-1	Long-term acceptance MHC-mismatched heart grafts	[101]
IL-2+ IL-15 Ig	Long-term acceptance of MHC-mismatched islets in diabetic NOD mice	[102]

CD4, CD40L, LFA-1 and ICAM-1, CD2, CD3, CD28, and CD45 can prevent or delay transplant rejection or autoimmunity (Tabs 1 and 2). In several experimental systems it has been demonstrated that allograft acceptance is a consequence of dominant tolerance that can resist the adoptive transfer of aggressive T cells [9], and even recruit some of these aggressive T cells into the regulatory pool – a phenomenon termed 'infectious tolerance' [10, 11]. It has also been shown that the tolerance state can be extended to third party antigens when they are provided in tissues co-expressing the tolerated antigens – an observation described as 'linked suppression' (Fig. 1) [12]. The transplantation of donor thymic epithelium has also been shown to be effective in inducing dominant tolerance, suggesting a contribution of Treg cells developed in the transplanted thymic allograft for active suppression of alloreactivity [13].

The use of altered peptide ligands (APL) is also effective in the induction of dominant tolerance to allografts in otherwise unmanipulated recipients [14, 15].

Table 2 - Monoclonal antibodies in experimental autoimmune diseases.

mAb	Animal/disease	Outcome	Ref.
CD2	Lewis rats(EAM)	Prevents onset	[103]
	Lewis rats (EAE)	Prevents onset	[104]
	BB/wor rats (type 1 diabetes)	Prevents onset	[105]
CD3	NOD mouse (Type 1 diabetes)	Ameliorates established disease	[106–109]
	DBA/1 mouse (CIA)	Delayed onset, reduced severity	[110]
CD4	NOD mouse (Type 1 diabetes)	Prevents onset	[111–114]
	NZB/NZW mouse (murine SLE)	Ameliorates established disease	[115]
	DBA/l (CIA)	Prevents onset and ameliorates established disease	[116]
CD30L	NOD mouse (type 1 diabetes)	Prevents onset	[117]
OX40L	SJL mouse (EAE)	Ameliorates established disease	[118]
CTLA4-Ig	BALB/c mouse (EAE)	Ameliorates established disease	[119]
	NZB/NZW mouse (SLE)	Prevents onset	[120–123]
	BXSB mouse (SLE)	Delayed onset, reduced severity	[124]
CD154	Marmoset monkey (EAE)	Prevents onset	[125]
	B6/A (C57BL/6×A/J)(AOD)	Prevents onset	[126]
	B10.BR mouse (EAU)	Ameliorates established disease	[127]
	SJL mouse (EAT)	Less severity	[128]
	NZB/NZW mouse (SLE)	Prevents onset	[129–132]
CD40	Marmoset monkey (EAE)	Prevents onset	[133]
CD137	DBA/1 (CIA)	Prevents onset	[134]
	NZB/NZW mouse (SLE)	Ameliorates established disease	[135]
	C.B-17 SCID (IBD)	Ameliorates established disease	[136, 137]
	C57BL/6 (EAE)	Ameliorates established disease	[138]
LFA-1	NOD mouse (Type 1 diabetes)	Prevents onset	[139]

EAE, experimental allergic encephalomyelitis; EAM, experimental autoimmune myelitis; AOD, autoimmune ovary disease; CIA, collagen-induced arthritis; EAU, experimental autoimmune uveoretinitis; IBD, inflammatory bowel disease; MS, multiple sclerosis; SLE, systemic lupus erythematosus.

Repetitive administration of an APL of the male antigen Dby leads to dominant tolerance associated with the development of Foxp3$^+$ Treg cells from anti-male TCR-transgenic T cells [16]. In addition, long-term exposure to a harmless anti-

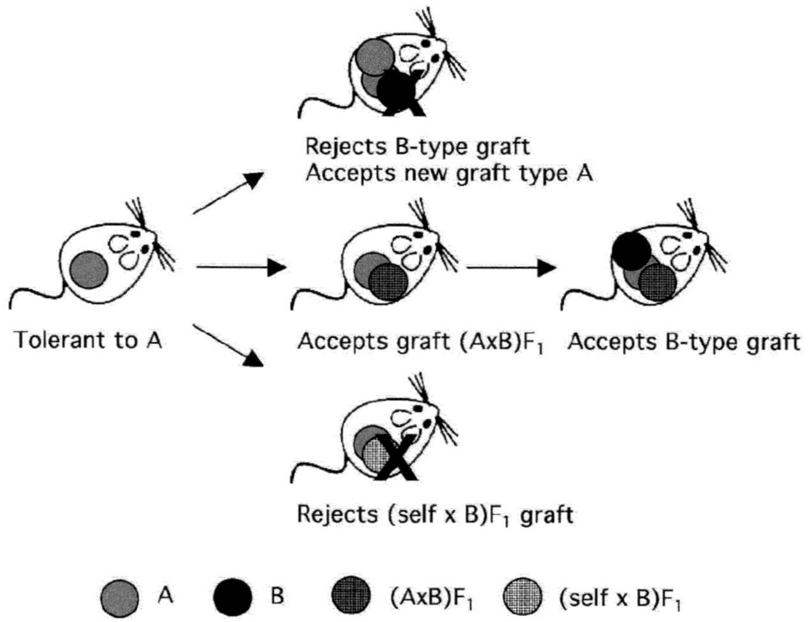

Figure 1

Linked suppression as a manifestation of dominant regulation.

Animals displaying Treg cell-mediated dominant tolerance to grafts of type A are fully competent to reject type B grafts; however, they readily accept $(A\times B)F_1$ grafts, where both sets of antigens (A and B) are present in the same cells. Furthermore, following the acceptance of $(A\times B)F_1$ grafts, the animals become tolerant of B-type grafts. Linked suppression is not observed to self antigens as tolerant mice reject $(\text{self}\times B)F_1$ grafts.

gen can also lead to the induction of dominant tolerance. This can be achieved by persistent exposure of TCR-transgenic T cells to their specific antigen provided by an osmotic pump [5], or following the adoptive transfer of TCR-transgenic T cells into T cell-deficient mice expressing the appropriate antigen [17]. Tolerance induction has also been reported following the administration of immature dendritic cells (DCs), or DCs in an altered state of maturation, such that their capacity to present antigen for full T cell activation is compromised [18, 19]. Such modulation of the DC stimulatory function can be achieved following the exposure to reagents [such as IL-10, 1α,25-dihydroxyvitamin D3 (vitD3), or LF 15-0195] that interfere with DC maturation [20–22]. The observation that apoptotic material delivery to DCs can result in immune tolerance [23] may also represent a way to achieve *in vivo* antigen presentation by immature DCs.

Such a dominant tolerance state is qualitatively different from tolerance induced to allografts following the development of mixed hematopoietic chimerism [24]. When donor bone marrow is used to induce tolerance to solid organs, the tolerance state is a consequence of the deletion of alloreactive cells and not the expansion of Treg cells [25]. Under these conditions, the tolerance state is recessive and is not able to suppress adoptively transferred aggressive T cells, or to prevent the rejection of linked antigens (linked suppression) [25].

Treg cells are required for the maintenance of dominant tolerance

Observations made over three decades ago concerning neonatal transplantation tolerance, have established that T cells could suppress responses to foreign proteins or allogeneic grafts, after adoptive transfer into irradiated secondary recipients [26, 27]. It was also shown that the development of autoimmune manifestations developing in neonatally thymectomized animals could be prevented by the adoptive transfer of thymocytes or splenocytes from normal syngeneic donors [28–30]. With the development of methods allowing specific depletion or sorting of T cell subsets, it became possible to further characterize the phenotype of the cells that could prevent the onset of autoimmune diseases or gut immunopathology upon adoptive transfer into susceptible animals and also in therapeutically induced tolerance. Initially, Sakaguchi and colleagues [31] identified the regulatory capacity among the $CD5^+$ T cells, while in experimentally induced tolerance regulatory activity was present among the $CD4^+$ T cells [32]. The $CD4^+$ T cells were further subdivided: first the regulatory activity was found to be within the $CD4^+CD45RC^{low}$ compartment in the rat or the $CD4^+CD45RB^{low}$ compartment in the mice, and later within the $CD4^+CD25^+$ subpopulation [33–35].

It has been shown that several of the strategies leading to the induction of tolerance, described in Tables 1 and 2, do so by inducing Treg cells [8, 36]. Furthermore, there is evidence that Treg cells maintaining transplantation tolerance can be found not only within the spleen and lymph nodes, but also infiltrating the tolerized allograft [37]. This observation suggests that part of the Treg cell action may be to impose a local state of immune privilege [38, 39]. In fact several reports have shown that Treg cells are involved in modulating indoleamine 2,3-dioxygenase (IDO) expression by DCs, and it is possible Treg cells may endow non-hematopoietic cells to produce anti-inflammatory genes like heme-oxygenase-1 [39].

The study of Treg cells in the maintenance of dominant tolerance had a significant boost following the identification of the transcription factor Foxp3 as a molecular marker of Treg cells able to impose a regulatory phenotype on non-regulatory T cells [40–42]. It became possible to use a molecular marker allowing a distinction between Treg cells and activated T cells, since an acute infection or TCR stimulation does not appear to trigger Foxp3 expression in mice [43]. However, recent reports have suggested that unlike in mice, Foxp3 may be transiently expressed in human activated T cells [44]. Although not all Treg cells express Foxp3 [45], absence of Foxp3 compromises immune tolerance, leading to severe autoimmunity both in animal models [46] and human patients [47].

There is now compelling evidence for the capacity of Foxp3$^+$ Treg cells in preventing autoimmunity, allergy, gut immunopathology, and transplant rejection [1]. However, Treg-mediated suppression may be associated with deleterious effects, namely inhibition of anti-tumor immune responses or protective responses against pathogens [48, 49]. A population of Foxp3$^+$ Treg cells critical for the prevention of autoimmunity is produced in the thymus [50]. Recent reports have also revealed that non-regulatory CD4$^+$ lymphocytes can be converted into Foxp3$^+$ Treg cells in peripheral tissues [51].

Therapeutic conversion of naive T cells into Treg cells

The study of peripheral Treg cell conversion was greatly facilitated with TCR-transgenic RAG-deficient mice. Such animals are unable to produce a functional TCR from their endogenous genes. As a consequence, all T cells exclusively express the transgenic TCR. Thymic development in the absence of an appropriate ligand for the transgenic TCR results in a population of mature T cells without Foxp3-expressing Treg cells [52]. Therefore, it becomes possible to examine the conversion of such T cells into Foxp3$^+$ Treg cells, in the absence of a possible contamination due to the expansion of pre-existing Treg cells.

Using male-antigen specific TCR-transgenic RAG$^{-/-}$ mice, it was possible to document the *de novo* induction of Treg cells following *in vitro* antigenic stimulation by DCs in the presence of non-depleting anti-CD4 mAb [7]. A similar conversion of non-regulatory T cells into Tregs was observed *in vivo* following transplantation of male skin grafts onto female TCR-transgenic mice treated with tolerogenic anti-CD4 mAbs [7]. The same mouse strain was used to demonstrate that *in vivo* exposure to the antigenic peptide or to an appropriate APL also leads to peripheral induction of Treg cells and dominant transplantation tolerance [8, 16]. Immature DCs or DCs treated with vitD3 (a reagent that prevents subsequent DC maturation), when adoptively transferred into male-specific TCR-transgenic female mice, also induce the peripheral generation of Foxp3$^+$ Treg cells and dominant tolerance to male skin grafts [53]. The peripheral induction of Treg cells, both *in vitro* and *in vivo*, seems to require TGF-β as it is abrogated in the presence of neutralizing anti-TGF-β mAbs [7, 53].

Chicken ovalbumin (OVA)-specific TCR-transgenic $RAG^{-/-}$ mice were also used to investigate extra-thymic conversion of $CD4^+$ T cells into Treg cells. In one of such studies, the exogenous addition of TGF-β to *in vitro* T cell cultures resulted in the conversion of some of the T cells into $Foxp3^+$ Treg cells [6]. Furthermore, the *in vitro* activation of $Foxp3^-$ T cells in cultures with a low concentration of the peptide also led to the induction of $Foxp3^+$ T cells [2]. Interestingly, B cells were shown to be more efficient than DCs in driving Treg conversion, presumably by their inability to provide full costimulatory signals [2].

The induction of Treg cells has also been reported following activation by DCs in an immature state – cells that, as described above, can lead to dominant tolerance *in vivo*. Such is the case with immature DCs, a DC subset expressing CD103, and IL-10- or vitamin D3-treated DCs. Furthermore, the combined effects of tryptophan starvation and tryptophan catabolites has been shown to down-modulate TCR zeta-chain, converting naive T cells into Treg cells [54].

Oral tolerance, induced in mice by exposure to OVA in the drinking water, was also shown to lead to the conversion of TCR-transgenic OVA-specific T cells into $Foxp3^+$ Treg cells [55]. Similarly, *in vivo* exposure of T cells to a low dose of persistent antigen also resulted in $Foxp3^+$ Treg induction [5]. Spontaneous conversion of non-regulatory T cells into $Foxp3^+$ Tregs was also claimed following adoptive transfer experiments of $CD4^+CD25^-$ T cells into congenic mice [56, 57]. However, in animals that are not TCR-transgenic $RAG^{-/-}$ it is always difficult to exclude a contribution of Treg expansion by some contaminating $Foxp3^+CD25^-$ cells [43].

A different population of Treg cells, $Foxp3^-$ and IL-10 producers, named Tr1 has been described [58]. Tr1 cells are peripherally induced by antigenic stimulation in an IL-10-rich environment. Although $Foxp3^+$ Treg cells are critical in preventing autoimmunity, Tr1 cells may become useful therapeutic tools for the suppression of immune pathology [59].

It should be noted that some of the experimental systems reported as leading to T cell anergy – a functional state in which T cells remain viable but unable to respond to optimal stimulation through both the TCR and costimulatory molecules – are similar to approaches leading to conversion of naive cells into Treg cells. Following initial studies by Lamb and colleagues [60] showing that T cell exposure to high doses of influenza virus hemagglutinin would lead to T cell anergy, it was shown that a similar state of T cell unresponsiveness could be achieved by antigen recognition in the absence of costimulation [61, 62], the use of APLs [63, 64], or direct presentation by activated rat or human T cells which express MHC class II molecules [65, 66].

T cell anergy was also reported *in vivo* in several animal models. Initially it was reported following transplantation tolerance induced with anti-CD4 mAbs [67, 68]. Other studies have described T cell anergy being induced by the injection of cells expressing the self-superantigen Mls-1a in mice [69], by aqueous peptide antigen administration in mice [70], in double transgenic mice for a TCR and its surrogate antigen [71, 72], and in oral tolerance [73].

Given the current tools available for the identification of Treg cells [74] it would be relevant to revisit these experimental systems and addressing the contribution of Treg cell induction.

The blind-spot of Treg cells

The expansion of Treg specific for tissue antigens offers a potential therapeutic tool for autoimmune diseases. Thus far, it has been possible to specifically expand TCR-transgenic Treg cells (BDC2.5) capable of preventing type 1 diabetes in NOD mice [75, 76], as well as cells with similar specificity from wild-type populations [77]. Interestingly, however, DCs were not competent APCs in expanding non-transgenic Treg, and this required additional TCR stimulation [76]. More recently, a DC subset was described as being especially effective in promoting the expansion of TCR-transgenic Treg cells, while unable to drive proliferation of endogenous Treg cells [78]. These observations suggest that, although the Treg repertoire is skewed towards self reactivity, it generally ignores DC-specific autoantigens, commanding consideration of this "blind-spot" when defining therapeutic strategies to expand Treg cells: antigens that are not normally available to thymic DCs may prove most efficient in mediating Treg expansion.

As Treg cells are positively selected by thymic epithelial cells (TECs) on the basis of self reactivity, they would systematically suppress protective immune responses unless their repertoire is devoid of recognition towards peripheral APCs. This may be achieved by negative selection of developing Tregs on thymic DCs, thus creating a "blind-spot" corresponding to DC-self antigens in the mature Treg cell repertoire [79].

This hypothesis is supported by observations related to linked suppression – a major feature of dominant tolerance, as described above. However, this phenomenon does not occur when the tolerated antigens are "natural self", in spite of the well-documented contribution of dominant regulation for the prevention of autoimmunity (Fig. 1). In fact, transplants from (self × third-party)F_1 donors are readily rejected [12]. This observation has been interpreted to indicate that self tolerance is essentially due to deletion of autoreactive T cell clones, with the putatively remaining Treg clones being too rare to exert efficient linked suppression on the T cells reactive to third-party antigens. Furthermore, the inability of DCs to expand rare Treg clones recognizing DC-associated antigens could imply that lack of linked suppression to self antigens reveals a qualitative, rather than quantitative, deficit. It has also been shown that microbial-derived "danger" signals, such as toll-like receptor ligands, can lead to a transitory inactivation of Treg cell function [80, 81]. However, this phenomenon alone cannot account for the data on linked suppression described above, or to the inability of DCs to expand endogenous Treg clones.

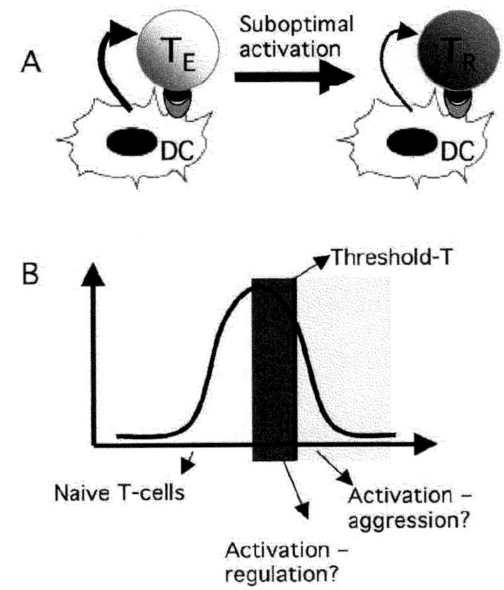

Figure 2
Suboptimal activation for peripheral induction of Treg cells.
Peripheral T cell activation may determine T cell fate depending on the stimulation intensity. (A) Reagents preventing optimal T cell activation (for instance, antibodies interfering with the formation of the immune synapse, TGF-β, or altered peptide ligands) can drive peripheral T cell commitment towards a Foxp3$^+$ regulatory phenotype. (B) The same TCR-transgenic T cell can be peripherally driven towards the regulatory or aggressive T cell pool depending on its activation reaching a critical threshold (threshold T).

Thus, the absence of Treg cell activity towards DCs and DC-associated antigens may be necessary to allow for immune responses following effective presentation of non-self antigens, such as microbial peptides, without the hindrance of linked suppression. In other words, to prevent the constant suppression of T cell immune responses towards foreign antigens presented by DC, the Treg cell repertoire ought to be "blind" to DC-self antigens [79].

Suboptimal activation for the peripheral induction of Treg cells

It is now generally accepted that thymic generation of Treg cells requires the recognition of antigen [52], with this requirement shaping the Treg cell TCR repertoire towards self recognition [82]. However, it appears that when thymic recognition reaches a certain threshold for the thymocyte activation, it results in the induction of apoptosis and negative selection [83]. The observations described in the previous section, together with knowledge of thymic Treg generation, led us to propose [51] that peripheral Treg induction probably mirrors the thymic events: if a T cell encounters antigen in an inflammatory environment supporting full activation, it will commit towards an aggressive phenotype appropriate to the initiation of a protective immune response. If a T cell interacts with the antigen in an environment conducive to suboptimal activation, it will differentiate towards a regulatory phenotype (Fig. 2).

The factors contributing to suboptimal activation: low concentration of the peptide, APL, immaturity of the DC, reagents interfering with the formation of the immune synapse, can therefore facilitate the conversion of naive T cells into Foxp3$^+$ Treg cells. The contribution of TGF-β for Treg conversion may be due to its role in increasing the threshold for T cell activation [140]. Our hypothesis is further supported by observations that mutations in T cell stimulatory components, such as Lck, may facilitate Treg cell development [84].

Conclusion

Ultimately, the realization that Treg cell conversion from non-regulatory precursors is a function of sub-optimal T cell activation may represent a shift in the current paradigm of instructive signals for peripheral commitment of effector T cells, towards a quantitative model: the cell fate may be a consequence of the intensity of activation rather than the presence or absence of certain molecules. Therefore, it may be quantitative differences in T cell activation status that governs the differentiation of Treg versus Th1/Th2/Th17 cells and thus tolerance *versus* immunity.

References

1 Sakaguchi S (2005) Naturally arising Foxp3-expressing CD25$^+$CD4$^+$ regulatory T cells in immunological tolerance to self and non-self. *Nat Immunol* 6: 345–352

2 Kim JM, Rudensky A (2006) The role of the transcription factor Foxp3 in the development of regulatory T cells. *Immunol Rev* 212: 86–98

3 Bluestone JA (2005) Regulatory T-cell therapy: is it ready for the clinic? *Nat Rev Immunol* 5: 343–349

4 Yamazaki S, Inaba K, Tarbell KV, Steinman RM (2006) Dendritic cells expand antigen-specific Foxp3$^+$ CD25$^+$ CD4$^+$ regulatory T cells including suppressors of alloreactivity. *Immunol Rev* 212: 314–329

5 Apostolou I, von Boehmer H (2004) *In vivo* instruction of suppressor commitment in naive T cells. *J Exp Med* 199: 1401–1408

6 Chen W, Jin W, Hardegen N, Lei KJ, Li L, Marinos N, McGrady G, Wahl SM (2003) Conversion of peripheral CD4$^+$CD25$^-$ naive T cells to CD4$^+$CD25$^+$ regulatory T cells by TGF-beta induction of transcription factor Foxp3. *J Exp Med* 198: 1875–1886

7 Cobbold SP, Castejon R, Adams E, Zelenika D, Graca L, Humm S, Waldmann H (2004) Induction of foxP3$^+$ regulatory T cells in the periphery of T cell receptor transgenic mice tolerized to transplants. *J Immunol* 172: 6003–6010

8 Waldmann H, Chen TC, Graca L, Adams E, Daley S, Cobbold S, Fairchild PJ (2006) Regulatory T cells in transplantation. *Semin Immunol* 18: 111–119

9 Qin SX, Wise M, Cobbold SP, Leong L, Kong YC, Parnes JR, Waldmann H (1990) Induction of tolerance in peripheral T cells with monoclonal antibodies. *Eur J Immunol* 20: 2737–2745

10 Graca L, Honey K, Adams E, Cobbold SP, Waldmann H (2000) Cutting edge: anti-CD154 therapeutic antibodies induce infectious transplantation tolerance. *J Immunol* 165: 4783–4786

11 Qin S, Cobbold SP, Pope H, Elliott J, Kioussis D, Davies J, Waldmann H (1993) "Infectious" transplantation tolerance. *Science* 259: 974–977

12 Davies JD, Leong LY, Mellor A, Cobbold SP, Waldmann H (1996) T cell suppression in transplantation tolerance through linked recognition. *J Immunol* 156: 3602–3607

13 Modigliani Y, Coutinho A, Pereira P, Le Douarin N, Thomas-Vaslin V, Burlen-Defranoux O, Salaun J, Bandeira A (1996) Establishment of tissue-specific tolerance is driven by regulatory T cells selected by thymic epithelium. *Eur J Immunol* 26: 1807–1815

14 Chai JG, James E, Dewchand H, Simpson E, Scott D (2004) Transplantation tolerance induced by intranasal administration of HY peptides. *Blood* 103: 3951–3959

15 James E, Scott D, Chai JG, Millrain M, Chandler P, Simpson E (2002) HY peptides modulate transplantation responses to skin allografts. *Int Immunol* 14: 1333–1342

16 Chen TC, Waldmann H, Fairchild PJ (2004) Induction of dominant transplantation tolerance by an altered peptide ligand of the male antigen Dby. *J Clin Invest* 113: 1754–1762

17 Chen TC, Cobbold SP, Fairchild PJ, Waldmann H (2004) Generation of anergic and regulatory T cells following prolonged exposure to a harmless antigen. *J Immunol* 172: 5900–5907

18 Mahnke K, Schmitt E, Bonifaz L, Enk AH, Jonuleit H (2002) Immature, but not inactive: the tolerogenic function of immature dendritic cells. *Immunol Cell Biol* 80: 477–483

19 Steinman RM, Hawiger D, Nussenzweig MC (2003) Tolerogenic dendritic cells. *Annu Rev Immunol* 21: 685–711

20 Steinbrink K, Wolfl M, Jonuleit H, Knop J, Enk AH (1997) Induction of tolerance by IL-10-treated dendritic cells. *J Immunol* 159: 4772–4780
21 Griffin MD, Lutz W, Phan VA, Bachman LA, McKean DJ, Kumar R (2001) Dendritic cell modulation by 1alpha,25 dihydroxyvitamin D3 and its analogs: a vitamin D receptor-dependent pathway that promotes a persistent state of immaturity *in vitro* and *in vivo*. *Proc Natl Acad Sci USA* 98: 6800–6805
22 Yang J, Bernier SM, Ichim TE, Li M, Xia X, Zhou D, Huang X, Strejan GH, White DJ, Zhong R et al (2003) LF15–0195 generates tolerogenic dendritic cells by suppression of NF-kappaB signaling through inhibition of IKK activity. *J Leukoc Biol* 74: 438–447
23 Liu K, Iyoda T, Saternus M, Kimura Y, Inaba K, Steinman RM (2002) Immune tolerance after delivery of dying cells to dendritic cells *in situ*. *J Exp Med* 196: 1091–1097
24 Wekerle T, Sykes M (2001) Mixed chimerism and transplantation tolerance. *Annu Rev Med* 52: 353–370
25 Graca L, Le Moine A, Lin CY, Fairchild PJ, Cobbold SP, Waldmann H (2004) Donor-specific transplantation tolerance: the paradoxical behavior of $CD4^{+}CD25^{+}$ T cells. *Proc Natl Acad Sci USA* 101: 10122–10126
26 Dorsch S, Roser B (1982) Suppressor cells in transplantation tolerance. I. Analysis of the suppressor status of neonatally and adoptively tolerized rats. *Transplantation* 33: 518–524
27 Gershon RK, Kondo K (1971) Infectious immunological tolerance. *Immunology* 21: 903–914
28 Kojima A, Taguchi O, Nishizuka Y (1980) Experimental production of possible autoimmune gastritis followed by macrocytic anemia in athymic nude mice. *Lab Invest* 42: 387–395
29 Kojima A, Tanaka-Kojima Y, Sakakura T, Nishizuka Y (1976) Prevention of postthymectomy autoimmune thyroiditis in mice. *Lab Invest* 34: 601–605
30 Nishizuka Y, Sakakura T (1969) Thymus and reproduction: sex-linked dysgenesia of the gonad after neonatal thymectomy in mice. *Science* 166: 753–755
31 Sakaguchi S, Fukuma K, Kuribayashi K, Masuda T (1985) Organ-specific autoimmune diseases induced in mice by elimination of T cell subset. I. Evidence for the active participation of T cells in natural self-tolerance; deficit of a T cell subset as a possible cause of autoimmune disease. *J Exp Med* 161: 72–87
32 Hall BM, Jelbart ME, Gurley KE, Dorsch SE (1985) Specific unresponsiveness in rats with prolonged cardiac allograft survival after treatment with cyclosporine. Mediation of specific suppression by T helper/inducer cells. *J Exp Med* 162: 1683–1694
33 Fowell D, Mason D (1993) Evidence that the T cell repertoire of normal rats contains cells with the potential to cause diabetes. Characterization of the $CD4^{+}$ T cell subset that inhibits this autoimmune potential. *J Exp Med* 177: 627–636
34 Powrie F, Mason D (1990) OX-22high $CD4^{+}$ T cells induce wasting disease with multiple organ pathology: prevention by the OX-22low subset. *J Exp Med* 172: 1701–1708
35 Sakaguchi S, Sakaguchi N, Asano M, Itoh M, Toda M (1995) Immunologic self-tolerance maintained by activated T cells expressing IL-2 receptor alpha-chains (CD25).

Breakdown of a single mechanism of self-tolerance causes various autoimmune diseases. *J Immunol* 155: 1151–1164

36 Chatenoud L (2006) Immune therapies of autoimmune diseases: are we approaching a real cure? *Curr Opin Immunol* 18: 710–717

37 Graca L, Cobbold SP, Waldmann H (2002) Identification of regulatory T cells in tolerated allografts. *J Exp Med* 195: 1641–1646

38 Cobbold SP, Adams E, Graca L, Daley S, Yates S, Paterson A, Robertson NJ, Nolan KF, Fairchild PJ, Waldmann H (2006) Immune privilege induced by regulatory T cells in transplantation tolerance. *Immunol Rev* 213: 239–255

39 Oliveira V, Agua-Doce A, Duarte J, Soares MP, Graca L (2006) Regulatory T cell maintenance of dominant tolerance: induction of tissue self-defense? *Transpl Immunol* 17: 7–10

40 Hori S, Nomura T, Sakaguchi S (2003) Control of regulatory T cell development by the transcription factor Foxp3. *Science* 299: 1057–1061

41 Fontenot JD, Gavin MA, Rudensky AY (2003) Foxp3 programs the development and function of $CD4^+CD25^+$ regulatory T cells. *Nat Immunol* 4: 330–336

42 Khattri R, Cox T, Yasayko SA, Ramsdell F (2003) An essential role for Scurfin in $CD4^+CD25^+$ T regulatory cells. *Nat Immunol* 4: 337–342

43 Fontenot JD, Rasmussen JP, Williams LM, Dooley JL, Farr AG, Rudensky AY (2005) Regulatory T cell lineage specification by the forkhead transcription factor foxp3. *Immunity* 22: 329–341

44 Wang J, Ioan-Facsinay A, van der Voort EI, Huizinga TW, Toes RE (2007) Transient expression of FOXP3 in human activated nonregulatory $CD4^+$ T cells. *Eur J Immunol* 37: 129–138

45 Levings MK, Roncarolo MG (2005) Phenotypic and functional differences between human $CD4^+CD25^+$ and type 1 regulatory T cells. *Curr Top Microbiol Immunol* 293: 303–326

46 Kim JM, Rasmussen JP, Rudensky AY (2007) Regulatory T cells prevent catastrophic autoimmunity throughout the lifespan of mice. *Nat Immunol* 8: 191–197

47 Chatila TA, Blaeser F, Ho N, Lederman HM, Voulgaropoulos C, Helms C, Bowcock AM (2000) JM2, encoding a fork head-related protein, is mutated in X-linked autoimmunity-allergic disregulation syndrome. *J Clin Invest* 106: R75–81

48 Suvas S, Rouse BT (2006) Treg control of antimicrobial T cell responses. *Curr Opin Immunol* 18: 344–348

49 Wang HY, Wang RF (2007) Regulatory T cells and cancer. *Curr Opin Immunol* 19: 217–223

50 Liston A, Rudensky AY (2007) Thymic development and peripheral homeostasis of regulatory T cells. *Curr Opin Immunol* 19: 176–185

51 Graca L, Chen TC, Le Moine A, Cobbold SP, Howie D, Waldmann H (2005) Dominant tolerance: activation thresholds for peripheral generation of regulatory T cells. *Trends Immunol* 26: 130–135

52 Jordan MS, Boesteanu A, Reed AJ, Petrone AL, Holenbeck AE, Lerman MA, Naji A,

Caton AJ (2001) Thymic selection of $CD4^+CD25^+$ regulatory T cells induced by an agonist self-peptide. *Nat Immunol* 2: 301–306

53 Yates SF, Paterson AM, Nolan KF, Cobbold SP, Saunders NJ, Waldmann H, Fairchild PJ (2007) Induction of regulatory T cells and dominant tolerance by dendritic cells incapable of full activation. *J Immunol* 179: 967–976

54 Fallarino F, Grohmann U, You S, McGrath BC, Cavener DR, Vacca C, Orabona C, Bianchi R, Belladonna ML, Volpi C et al (2006) The combined effects of tryptophan starvation and tryptophan catabolites down-regulate T cell receptor zeta-chain and induce a regulatory phenotype in naive T cells. *J Immunol* 176: 6752–6761

55 Mucida D, Kutchukhidze N, Erazo A, Russo M, Lafaille JJ, Curotto de Lafaille MA (2005) Oral tolerance in the absence of naturally occurring Tregs. *J Clin Invest* 115: 1923–1933

56 Curotto de Lafaille MA, Lino AC, Kutchukhidze N, Lafaille JJ (2004) $CD25^-$ T cells generate $CD25^+Foxp3^+$ regulatory T cells by peripheral expansion. *J Immunol* 173: 7259–7268

57 Liang S, Alard P, Zhao Y, Parnell S, Clark SL, Kosiewicz MM (2005) Conversion of $CD4^+$ $CD25^-$ cells into $CD4^+$ $CD25^+$ regulatory T cells *in vivo* requires B7 costimulation, but not the thymus. *J Exp Med* 201: 127–137

58 Roncarolo MG, Gregori S, Battaglia M, Bacchetta R, Fleischhauer K, Levings MK (2006) Interleukin-10–secreting type 1 regulatory T cells in rodents and humans. *Immunol Rev* 212: 28–50

59 Battaglia M, Gregori S, Bacchetta R, Roncarolo MG (2006) Tr1 cells: from discovery to their clinical application. *Semin Immunol* 18: 120–127

60 Lamb JR, Skidmore BJ, Green N, Chiller JM, Feldmann M (1983) Induction of tolerance in influenza virus-immune T lymphocyte clones with synthetic peptides of influenza hemagglutinin. *J Exp Med* 157: 1434–1447

61 Jenkins MK, Schwartz RH (1987) Antigen presentation by chemically modified splenocytes induces antigen-specific T cell unresponsiveness *in vitro* and *in vivo*. *J Exp Med* 165: 302–319

62 Quill H, Schwartz RH (1987) Stimulation of normal inducer T cell clones with antigen presented by purified Ia molecules in planar lipid membranes: specific induction of a long-lived state of proliferative nonresponsiveness. *J Immunol* 138: 3704–3712

63 Sloan-Lancaster J, Evavold BD, Allen PM (1993) Induction of T-cell anergy by altered T-cell-receptor ligand on live antigen-presenting cells. *Nature* 363: 156–159

64 Sloan-Lancaster J, Evavold BD, Allen PM (1994) Th2 cell clonal anergy as a consequence of partial activation. *J Exp Med* 180: 1195–1205

65 Lombardi G, Hargreaves R, Sidhu S, Imami N, Lightstone L, Fuller-Espie S, Ritter M, Robinson P, Tarnok A, Lechler R (1996) Antigen presentation by T cells inhibits IL-2 production and induces IL-4 release due to altered cognate signals. *J Immunol* 156: 2769–2775

66 Taams LS, van Rensen AJ, Poelen MC, van Els CA, Besseling AC, Wagenaar JP, van

Eden W, Wauben MH (1998) Anergic T cells actively suppress T cell responses *via* the antigen-presenting cell. *Eur J Immunol* 28: 2902–2912

67 Qin SX, Cobbold S, Benjamin R, Waldmann H (1989) Induction of classical transplantation tolerance in the adult. *J Exp Med* 169: 779–794

68 Alters SE, Shizuru JA, Ackerman J, Grossman D, Seydel KB, Fathman CG (1991) Anti-CD4 mediates clonal anergy during transplantation tolerance induction. *J Exp Med* 173: 491–494

69 Rammensee HG, Kroschewski R, Frangoulis B (1989) Clonal anergy induced in mature V beta 6$^+$ T lymphocytes on immunizing Mls-1b mice with Mls-1a expressing cells. *Nature* 339: 541–544

70 Burstein HJ, Shea CM, Abbas AK (1992) Aqueous antigens induce *in vivo* tolerance selectively in IL-2- and IFN-gamma-producing (Th1) cells. *J Immunol* 148: 3687–3691

71 Jordan MS, Riley MP, von Boehmer H, Caton AJ (2000) Anergy and suppression regulate CD4($^+$) T cell responses to a self peptide. *Eur J Immunol* 30: 136–144

72 Schonrich G, Momburg F, Hammerling GJ, Arnold B (1992) Anergy induced by thymic medullary epithelium. *Eur J Immunol* 22: 1687–1691

73 Chen Y, Kuchroo VK, Inobe J, Hafler DA, Weiner HL (1994) Regulatory T cell clones induced by oral tolerance: suppression of autoimmune encephalomyelitis. *Science* 265: 1237–1240

74 Graca L (2005) New tools to identify regulatory T cells. *Eur J Immunol* 35: 1678–1680

75 Tang Q, Henriksen KJ, Bi M, Finger EB, Szot G, Ye J, Masteller EL, McDevitt H, Bonyhadi M, Bluestone JA (2004) *In vitro*-expanded antigen-specific regulatory T cells suppress autoimmune diabetes. *J Exp Med* 199: 1455–1465

76 Tarbell KV, Yamazaki S, Olson K, Toy P, Steinman RM (2004) CD25$^+$ CD4$^+$ T cells, expanded with dendritic cells presenting a single autoantigenic peptide, suppress autoimmune diabetes. *J Exp Med* 199: 1467–1477

77 Masteller EL, Warner MR, Tang Q, Tarbell KV, McDevitt H, Bluestone JA (2005) Expansion of functional endogenous antigen-specific CD4$^+$CD25$^+$ regulatory T cells from nonobese diabetic mice. *J Immunol* 175: 3053–3059

78 Fisson S, Djelti F, Trenado A, Billiard F, Liblau R, Klatzmann D, Cohen JL, Salomon BL (2006) Therapeutic potential of self-antigen-specific CD4$^+$ CD25$^+$ regulatory T cells selected *in vitro* from a polyclonal repertoire. *Eur J Immunol* 36: 817–827

79 Graca L, Silva-Santos B, Coutinho A (2006) The blind-spot of regulatory T cells. *Eur J Immunol* 36: 802–805

80 Pasare C, Medzhitov R (2003) Toll pathway-dependent blockade of CD4$^+$CD25$^+$ T cell-mediated suppression by dendritic cells. *Science* 299: 1033–1036

81 Sutmuller RP, den Brok MH, Kramer M, Bennink EJ, Toonen LW, Kullberg BJ, Joosten LA, Akira S, Netea MG, Adema GJ (2006) Toll-like receptor 2 controls expansion and function of regulatory T cells. *J Clin Invest* 116: 485–494

82 Hsieh CS, Liang Y, Tyznik AJ, Self SG, Liggitt D, Rudensky AY (2004) Recognition

of the peripheral self by naturally arising CD25⁺ CD4⁺ T cell receptors. *Immunity* 21: 267–277

83 Lerman MA, Larkin J 3rd, Cozzo C, Jordan MS, Caton AJ (2004) CD4⁺ CD25⁺ regulatory T cell repertoire formation in response to varying expression of a neo-self-antigen. *J Immunol* 173: 236–244

84 Veldhoen M, Magee AI, Penha-Goncalves MN, Stockinger B (2005) Transduction of naive CD4 T cells with kinase-deficient Lck-HIV-Tat fusion protein dampens T cell activation and provokes a switch to regulatory function. *Eur J Immunol* 35: 207–216

85 Chavin KD, Lau HT, Bromberg JS (1992) Prolongation of allograft and xenograft survival in mice by anti-CD2 monoclonal antibodies. *Transplantation* 54: 286–291

86 Nicolls MR, Aversa GG, Pearce NW, Spinelli A, Berger MF, Gurley KE, Hall BM (1993) Induction of long-term specific tolerance to allografts in rats by therapy with an anti-CD3-like monoclonal antibody. *Transplantation* 55: 459–468

87 Chavin KD, Qin L, Lin J, Yagita H, Bromberg JS (1993) Combined anti-CD2 and anti-CD3 receptor monoclonal antibodies induce donor-specific tolerance in a cardiac transplant model. *J Immunol* 151: 7249–7259

88 Winsor-Hines D, Merrill C, O'Mahony M, Rao PE, Cobbold SP, Waldmann H, Ringler DJ, Ponath PD (2004) Induction of immunological tolerance/hyporesponsiveness in baboons with a nondepleting CD4 antibody. *J Immunol* 173: 4715–4723

89 Chen ZK, Cobbold SP, Waldmann H, Metcalfe S (1996) Amplification of natural regulatory immune mechanisms for transplantation tolerance. *Transplantation* 62: 1200–1206

90 Lenschow DJ, Zeng Y, Thistlethwaite JR, Montag A, Brady W, Gibson MG, Linsley PS, Bluestone JA (1992) Long-term survival of xenogeneic pancreatic islet grafts induced by CTLA4lg. *Science* 257: 789–792

91 Lazarovits AI, Poppema S, Zhang Z, Khandaker M, Le Feuvre CE, Singhal SK, Garcia BM, Ogasa N, Jevnikar AM, White MH et al (1996) Prevention and reversal of renal allograft rejection by antibody against CD45RB. *Nature* 380: 717–720

92 Basadonna GP, Auersvald L, Khuong CQ, Zheng XX, Kashio N, Zekzer D, Minozzo M, Qian H, Visser L, Diepstra A et al (1998) Antibody-mediated targeting of CD45 isoforms: a novel immunotherapeutic strategy. *Proc Natl Acad Sci USA* 95: 3821–3826

93 Shoda LK, Young DL, Ramanujan S, Whiting CC, Atkinson MA, Bluestone JA, Eisenbarth GS, Mathis D, Rossini AA, Campbell SE et al (2005) A comprehensive review of interventions in the NOD mouse and implications for translation. *Immunity* 23: 115–126

94 Curry AJ, Chikwe J, Smith XG, Cai M, Schwarz H, Bradley JA, Bolton EM (2004) OX40 (CD134) blockade inhibits the co-stimulatory cascade and promotes heart allograft survival. *Transplantation* 78: 807–814

95 Wu Z, Bensinger SJ, Zhang J, Chen C, Yuan X, Huang X, Markmann JF, Kassaee A, Rosengard BR, Hancock WW et al (2004) Homeostatic proliferation is a barrier to transplantation tolerance. *Nat Med* 10: 87–92

96 Parker DC, Greiner DL, Phillips NE, Appel MC, Steele AW, Durie FH, Noelle RJ,

Mordes JP, Rossini AA (1995) Survival of mouse pancreatic islet allografts in recipients treated with allogeneic small lymphocytes and antibody to CD40 ligand. *Proc Natl Acad Sci USA* 92: 9560–9564

97 Honey K, Cobbold SP, Waldmann H (1999) CD40 ligand blockade induces CD4[+] T cell tolerance and linked suppression. *J Immunol* 163: 4805–4810

98 Larsen CP, Elwood ET, Alexander DZ, Ritchie SC, Hendrix R, Tucker-Burden C, Cho HR, Aruffo A, Hollenbaugh D, Linsley PS et al (1996) Long-term acceptance of skin and cardiac allografts after blocking CD40 and CD28 pathways. *Nature* 381: 434–438

99 Korganow AS, Ji H, Mangialaio S, Duchatelle V, Pelanda R, Martin T, Degott C, Kikutani H, Rajewsky K, Pasquali JL et al (1999) From systemic T cell self-reactivity to organ-specific autoimmune disease *via* immunoglobulins. *Immunity* 10: 451–461

100 Benjamin RJ, Qin SX, Wise MP, Cobbold SP, Waldmann H (1988) Mechanisms of monoclonal antibody-facilitated tolerance induction: a possible role for the CD4 (L3T4) and CD11a (LFA-1) molecules in self-non-self discrimination. *Eur J Immunol* 18: 1079–1088

101 Isobe M, Yagita H, Okumura K, Ihara A (1992) Specific acceptance of cardiac allograft after treatment with antibodies to ICAM-1 and LFA-1. *Science* 255: 1125–1127

102 Zheng XX, Sanchez-Fueyo A, Sho M, Domenig C, Sayegh MH, Strom TB (2003) Favorably tipping the balance between cytopathic and regulatory T cells to create transplantation tolerance. *Immunity* 19: 503–514

103 Inomata T, Watanabe T, Haga M, Hirahara H, Abo T, Okura Y, Hanawa H, Kodama M, Izumi T (2000) Anti-CD2 monoclonal antibodies prevent the induction of experimental autoimmune myocarditis. *Jpn Heart J* 41: 507–517

104 Jung S, Toyka K, Hartung HP (1995) Suppression of experimental autoimmune encephalomyelitis in Lewis rats by antibodies against CD2. *Eur J Immunol* 25: 1391–1398

105 Barlow AK, Like AA (1992) Anti-CD2 monoclonal antibodies prevent spontaneous and adoptive transfer of diabetes in the BB/Wor rat. *Am J Pathol* 141: 1043–1051

106 Chatenoud L, Thervet E, Primo J, Bach JF (1994) Anti-CD3 antibody induces long-term remission of overt autoimmunity in nonobese diabetic mice. *Proc Natl Acad Sci USA* 91: 123–127

107 Chatenoud L, Primo J, Bach JF (1997) CD3 antibody-induced dominant self tolerance in overtly diabetic NOD mice. *J Immunol* 158: 2947–2954

108 von Herrath MG, Coon B, Wolfe T, Chatenoud L (2002) Nonmitogenic CD3 antibody reverses virally induced (rat insulin promoter-lymphocytic choriomeningitis virus) autoimmune diabetes without impeding viral clearance. *J Immunol* 168: 933–941

109 Chatenoud L (2003) CD3-specific antibody-induced active tolerance: from bench to bedside. *Nat Rev Immunol* 3: 123–132

110 Hughes C, Wolos JA, Giannini EH, Hirsch R (1994) Induction of T helper cell hyporesponsiveness in an experimental model of autoimmunity by using nonmitogenic anti-CD3 monoclonal antibody. *J Immunol* 153: 3319–3325

111 Hahn HJ, Kuttler B, Laube F, Emmrich F (1993) Anti-CD4 therapy in recent-onset IDDM. *Diabetes Metab Rev* 9: 323–328

112 Phillips JM, Harach SZ, Parish NM, Fehervari Z, Haskins K, Cooke A (2000) Nondepleting anti-CD4 has an immediate action on diabetogenic effector cells, halting their destruction of pancreatic beta cells. *J Immunol* 165: 1949–1955
113 Guo Z, Wu T, Kirchhof N, Mital D, Williams JW, Azuma M, Sutherland DE, Hering BJ (2001) Immunotherapy with nondepleting anti-CD4 monoclonal antibodies but not CD28 antagonists protects islet graft in spontaneously diabetic nod mice from autoimmune destruction and allogeneic and xenogeneic graft rejection. *Transplantation* 71: 1656–1665
114 Makhlouf L, Grey ST, Dong V, Csizmadia E, Arvelo MB, Auchincloss H Jr., Ferran C, Sayegh MH (2004) Depleting anti-CD4 monoclonal antibody cures new-onset diabetes, prevents recurrent autoimmune diabetes, and delays allograft rejection in nonobese diabetic mice. *Transplantation* 77: 990–997
115 Adachi Y, Inaba M, Sugihara A, Koshiji M, Sugiura K, Amoh Y, Mori S, Kamiya T, Genba H, Ikehara S (1998) Effects of administration of monoclonal antibodies (anti-CD4 or anti-CD8) on the development of autoimmune diseases in (NZW × BXSB)F1 mice. *Immunobiology* 198: 451–464
116 Mauri C, Chu CQ, Woodrow D, Mori L, Londei M (1997) Treatment of a newly established transgenic model of chronic arthritis with nondepleting anti-CD4 monoclonal antibody. *J Immunol* 159: 5032–5041
117 Chakrabarty S, Nagata M, Yasuda H, Wen L, Nakayama M, Chowdhury SA, Yamada K, Jin Z, Kotani R, Moriyama H et al (2003) Critical roles of CD30/CD30L interactions in murine autoimmune diabetes. *Clin Exp Immunol* 133: 318–325
118 Nohara C, Akiba H, Nakajima A, Inoue A, Koh CS, Ohshima H, Yagita H, Mizuno Y, Okumura K (2001) Amelioration of experimental autoimmune encephalomyelitis with anti-OX40 ligand monoclonal antibody: a critical role for OX40 ligand in migration, but not development, of pathogenic T cells. *J Immunol* 166: 2108–2115
119 Croxford JL, O'Neill JK, Ali RR, Browne K, Byrnes AP, Dallman MJ, Wood MJ, Fedlmann M, Baker D (1998) Local gene therapy with CTLA4-immunoglobulin fusion protein in experimental allergic encephalomyelitis. *Eur J Immunol* 28: 3904–3916
120 Finck BK, Linsley PS, Wofsy D (1994) Treatment of murine lupus with CTLA4Ig. *Science* 265: 1225–1227
121 Mihara M, Tan I, Chuzhin Y, Reddy B, Budhai L, Holzer A, Gu Y, Davidson A (2000) CTLA4Ig inhibits T cell-dependent B-cell maturation in murine systemic lupus erythematosus. *J Clin Invest* 106: 91–101
122 Daikh DI, Wofsy D (2001) Cutting edge: reversal of murine lupus nephritis with CTLA4Ig and cyclophosphamide. *J Immunol* 166: 2913–2916
123 Wang X, Huang W, Mihara M, Sinha J, Davidson A (2002) Mechanism of action of combined short-term CTLA4Ig and anti-CD40 ligand in murine systemic lupus erythematosus. *J Immunol* 168: 2046–2053
124 Chu EB, Hobbs MV, Wilson CB, Romball CG, Linsley PS, Weigle WO (1996) Intervention of $CD4^+$ cell subset shifts and autoimmunity in the BXSB mouse by murine CTLA4Ig. *J Immunol* 156: 1262–1268

125 Boon L, Brok HP, Bauer J, Ortiz-Buijsse A, Schellekens MM, Ramdien-Murli S, Blezer E, van Meurs M, Ceuppens J, de Boer M et al (2001) Prevention of experimental autoimmune encephalomyelitis in the common marmoset (*Callithrix jacchus*) using a chimeric antagonist monoclonal antibody against human CD40 is associated with altered B cell responses. *J Immunol* 167: 2942–2949
126 Sharp C, Thompson C, Samy ET, Noelle R, Tung KS (2003) CD40 ligand in pathogenesis of autoimmune ovarian disease of day 3-thymectomized mice: implication for CD40 ligand antibody therapy. *J Immunol* 170: 1667–1674
127 Namba K, Ogasawara K, Kitaichi N, Morohashi T, Sasamoto Y, Kotake S, Matsuda H, Iwabuchi K, Iwabuchi C, Ohno S et al (2000) Amelioration of experimental autoimmune uveoretinitis by pretreatment with a pathogenic peptide in liposome and anti-CD40 ligand monoclonal antibody. *J Immunol* 165: 2962–2969
128 Carayanniotis G, Masters SR, Noelle RJ (1997) Suppression of murine thyroiditis *via* blockade of the CD40-CD40L interaction. *Immunology* 90: 421–426
129 Early GS, Zhao W, Burns CM (1996) Anti-CD40 ligand antibody treatment prevents the development of lupus-like nephritis in a subset of New Zealand black × New Zealand white mice. Response correlates with the absence of an anti-antibody response. *J Immunol* 157: 3159–3164
130 Zandman-Goddard G, Shoenfeld Y (2000) Novel approaches to therapy for systemic lupus erythematosus. *Eur J Intern Med* 11: 130–134
131 Kalled SL, Cutler AH, Burkly LC (2001) Apoptosis and altered dendritic cell homeostasis in lupus nephritis are limited by anti-CD154 treatment. *J Immunol* 167: 1740–1747
132 Quezada SA, Eckert M, Adeyi OA, Schned AR, Noelle RJ, Burns CM (2003) Distinct mechanisms of action of anti-CD154 in early *versus* late treatment of murine lupus nephritis. *Arthritis Rheum* 48: 2541–2554
133 t Hart BA, Blezer EL, Brok HP, Boon L, de Boer M, Bauer J, Laman JD (2005) Treatment with chimeric anti-human CD40 antibody suppresses MRI-detectable inflammation and enlargement of pre-existing brain lesions in common marmosets affected by MOG-induced EAE. *J Neuroimmunol* 163: 31–39
134 Foell JL, Diez-Mendiondo BI, Diez OH, Holzer U, Ruck P, Bapat AS, Hoffmann MK, Mittler RS, Dannecker GE (2004) Engagement of the CD137 (4-1BB) costimulatory molecule inhibits and reverses the autoimmune process in collagen-induced arthritis and establishes lasting disease resistance. *Immunology* 113: 89–98
135 Foell J, Strahotin S, O'Neil SP, McCausland MM, Suwyn C, Haber M, Chander PN, Bapat AS, Yan XJ, Chiorazzi N et al (2003) CD137 costimulatory T cell receptor engagement reverses acute disease in lupus-prone NZB × NZW F1 mice. *J Clin Invest* 111: 1505–1518
136 Malmstrom V, Shipton D, Singh B, Al-Shamkhani A, Puklavec MJ, Barclay AN, Powrie F (2001) CD134L expression on dendritic cells in the mesenteric lymph nodes drives colitis in T cell-restored SCID mice. *J Immunol* 166: 6972–6981
137 Lee J, Lee EN, Kim EY, Park HJ, Chang CY, Jung DY, Choi SY, Lee SK, Lee KW, Kwon

GY et al (2005) Administration of agonistic anti-4-1BB monoclonal antibody leads to the amelioration of inflammatory bowel disease. *Immunol Lett* 101: 210–216

138 Sun Y, Lin X, Chen HM, Wu Q, Subudhi SK, Chen L, Fu YX (2002) Administration of agonistic anti-4–1BB monoclonal antibody leads to the amelioration of experimental autoimmune encephalomyelitis. *J Immunol* 168: 1457–1465

139 Moriyama H, Yokono K, Amano K, Nagata M, Hasegawa Y, Okamoto N, Tsukamoto K, Miki M, Yoneda R, Yagi N et al (1996) Induction of tolerance in murine autoimmune diabetes by transient blockade of leukocyte function-associated antigen-1/intercellular adhesion molecule-1 pathway. *J Immunol* 157: 3737–3743

140 Bommireddy R, Saxena V, Ormsby I, Yin M, Boivin GP, Babcock GF, Singh PR, Doetschmann T (2003) TGF-beta 1 regulates lymphocyte homeostasis by preventing activation and subsequent apoptosis of peripheral lymphocytes. *J Immunol* 170: 4612–4622

Infiltrating the immunological synapse: prospects for the use of altered peptide ligands for the treatment of immune pathology

Paul J. Fairchild

University of Oxford, Sir William Dunn School of Pathology, South Parks Road, Oxford, OX1 3RE, UK

Altered peptide ligands: A brief history

For many years, T cell activation was thought to lack finesse, engagement of the T cell receptor (TCR) resulting in the full spectrum of responses and changes in gene expression associated with activation. The inflexibility of such an all-or-nothing response provided few strategies for intervention in ongoing immune pathology, the most effective approach being the prevention of TCR engagement by MHC blockade. The use of surrogate peptides to compete for binding with the pathogenic epitope relied solely on their high affinity for the relevant MHC restriction element rather than any similarity in sequence to the pathogenic epitope [1]. The first description of altered peptide ligands (APL) therefore served as a turning point in understanding of T cells and their mode of antigen recognition, and suggested alternative approaches to immune intervention that relied on use of ligands that differ almost imperceptibly from the wild-type epitope, guaranteeing their recognition by the relevant T cells.

One of the first demonstrations of the inadequacy of conventional views of T cell activation was provided by the studies of De Magistris and colleagues [2]. Working with a human T cell clone specific for the influenza hemagglutinin epitope $HA_{307-319}$ in the context of HLA-DR1, analogues were constructed that were capable of binding to the restriction element but which failed to secure T cell activation. That these peptides were capable of engaging the TCR was, however, inferred from the way in which they potently inhibited responses to the wild-type epitope, even under circumstances in which competition for binding to HLA-DR1 had been minimized [2]. These findings strongly suggested that the TCR is fully susceptible to antagonism, in much the same way as other classical receptor-ligand interactions. Subsequent studies revealed antagonism to be a universal phenomenon that applied equally to MHC class I-restricted epitopes [3] and established a clear correlation between the similarity of an antagonist to its wild-type counterpart and its evident potency *in*

The Immune Synapse as a Novel Target for Therapy, edited by Luis Graca

vitro [4]. This rule was subsequently generalized by demonstrating that conservative changes in the secondary TCR contact sites of an epitope resulted in the most effective antagonists, targeting of the primary contact sites precluding all recognition by antigen-specific T cells, and resulting, instead, in a null phenotype [5].

A seminal study by Allen and colleagues (reviewed in [6]) extended these findings by demonstrating how T cell activation could best be conceptualized as a continuum with full agonism at one extreme, antagonism at the other and all manner of partial responses in between. Their conclusions were based on a molecular dissection of an epitope of hemoglobin from mice expressing the Hb^d allotype, which differs from Hb^s at residues 72, 73 and 76. The epitope $Hb^d_{64–76}$, which encompasses all three of these residues, is, therefore, sufficiently immunogenic in mice of the Hb^s allotype to enable the derivation of antigen-specific Th1 and Th2 cell clones. Interestingly, substitution of Glu at position 73 for Asp created an APL that, although fully recognized by the Th2 clone as evidenced by secretion of IL-4, failed to elicit a proliferative response [7]. Furthermore, the same peptide analogue could actively stimulate the cytolytic activity of a Th1 clone but in the absence of either proliferation or cytokine production [8]. Together with similar results obtained from MHC class I-restricted $CD8^+$ T cells [9], these findings provided the first compelling demonstration that what were once considered different manifestations of the same physiological response to antigen, were, in reality, distinct responses that could be physically uncoupled from one another.

While of obvious academic interest, these findings were afforded greater significance by studies in which partial agonists were shown to actively moderate the outcome of subsequent encounters with antigen. Working with a Th1 cell clone specific for $Hb^d_{64–76}$, Sloan-Lancaster and co-workers [10] demonstrated that exposure to an APL bearing Ser at position 70, resulted in profound anergy upon subsequent exposure to the wild-type epitope, results that could be replicated with a Th2 clone challenged with the Asp73 analogue of the same peptide [11]. Even more surprisingly, an APL of moth cytochrome c (MCC), which displayed low affinity for MHC class II, skewed naïve TCR-transgenic T cells towards a Th2 phenotype in contrast to the wild-type epitope, which provoked a predominantly Th1 response [12]. Given that immune deviation has been widely pursued as an approach to sublimating pathogenic immune responses *in vivo*, these results suggested that APL may prove more than just an immunological curiosity, offering opportunities to influence the decision making process at the very heart of the immune response in a manner conducive to the treatment of ongoing immune pathology.

Mode of action of APL

Although 15 years have now passed since their first description, the molecular mechanisms by which APL influence T cell responses remain a matter of some debate. At least three hypotheses have been proposed for the phenomenon of antagonism, each of which has accrued a body of supporting evidence (reviewed in [13]). The kinetic discrimination model is, for instance, based on the assumption that complexes between MHC determinants and antagonist peptides display an inherently lower affinity for the TCR than their wild-type counterparts, resulting in more rapid dissociation, incompatible with productive signaling [4]. This hypothesis has gained support from direct measurements of the affinity of a soluble TCR for complexes between the murine MHC class II determinant, H-2E^k, and analogues of MCC: for all three APL studied, the affinity was shown to be between 10- and 50-fold lower than that of the native epitope [14]. Nevertheless, such a notion has been challenged by observations that antagonists may function at concentrations one thousandth of those required for MHC blockade: quite how such limiting concentrations of an antagonist could interfere with responses to a surfeit of wild-type peptide, whose higher affinity would favor binding to the MHC, is difficult to envisage. Furthermore, the description of potent antagonists that display tangibly higher affinity for MHC molecules than the epitope on which they were modeled [15], suggests that issues of affinity may ultimately prove irrelevant.

An alternative hypothesis has focused on the likely changes in the three-dimensional structure of a peptide-MHC complex following mutation of a single amino acid residue. The corresponding conformational changes in the TCR upon ligation might be sufficient to affect the accessibility of protein tyrosine kinases to their substrates, for which evidence has begun to accumulate [16–19]. Nevertheless, an elegant study of a TCR bound to complexes between its cognate peptide and HLA-A2, showed the crystal structure to be almost identical to those of three APL, with only minor changes at the interface being detected, necessary to accommodate the amino acid substitution [20]. Critically, despite the similarity in their tertiary structure when bound to HLA-A2, APL solicited quite distinct responses from an antigen-specific T cell clone, compared to the wild-type peptide.

In the light of these deficiencies, the most compelling model of antagonism favors the capacity of some APL to elicit a dominant negative signaling cascade. Evidence in support of this concept comes from studies of naïve T cells expressing two TCR of distinct specificity. Robertson and co-workers [21] crossed two TCR-transgenic mouse strains, one with specificity for an epitope of hen egg lysozyme (HEL_{48-62}) in the context of H-2A^k, the other specific for complexes between MCC_{88-103} and H-2E^k. Interestingly, antagonist peptides based on the sequence of HEL, inhibited responses delivered by full agonists of MCC through the alternative TCR. Given that the two species of peptide were presented by different MHC molecules to distinct TCR, competition for binding could be operationally excluded as an explanation, strongly suggesting the transmission of a dominant negative signal. Although a second study of cytolytic activity among T cells co-expressing two unrelated MHC class I-restricted TCR failed to reach the same conclusions [22], subsequent experiments revealed the importance of selecting appropriate readouts for cross-antagonism, proliferation of $CD8^+$ T cells providing results fully compatible with a dominant negative signal [23].

Irrespective of the mechanisms of antagonism, it has recently become clear that the mode of action of partial agonists extends well beyond immediate events occurring at the immunological synapse to downstream signaling pathways (reviewed in [5]). Working in the Hb^d_{64-76} system, in which APL-induced anergy had first been described, Sloan-Lancaster and colleagues [24] investigated the intracellular signaling events responsible for the unresponsive state. While the native epitope stimulated the conventional pattern of CD3ζ phosphorylation, the Ser70 analogue, known to induce anergy among a Th1 cell clone, generated a unique pattern of tyrosine phosphorylation and failed to recruit the syk family kinase, ZAP-70. These findings have since been verified by other groups using both class I and class II-restricted T cells, strongly suggesting a general paradigm [9, 25]. That subtle changes in phosphorylation patterns can have profound downstream effects was demonstrated by Singh et al., who investigated an analogue of the 83–99 epitope of myelin basic protein (MBP). This APL had been identified by virtue of its ability to induce a phenotypic switch from a Th1 to a Th2 pattern of cytokine release by an antigen-specific T cell clone, derived from patients with multiple sclerosis. Intriguingly, the observed switch involved the differential activation of ERK, JNK and p38 MAPK, a Th1 phenotype correlating with enhanced JNK and p38 activity, while a Th2 switch was strongly associated with an up-regulation of ERK at the expense of JNK and p38 [26]. These results have important implications, not only for our understanding of the mechanisms underlying immune deviation, but for the prospects they offer for the rational design of APL capable of interfering in ongoing immune pathology in man. To date, most interest has focused on autoimmunity since the etiology of many autoimmune diseases is now well understood and both the autoantigenic epitopes and their MHC restriction elements well defined in many animal models and their human counterparts.

Immune intervention in autoimmunity

While the possibility of selectively antagonizing the activity of the very T cells responsible for unfettered autoimmunity remains a compelling prospect, which has enjoyed some success in animal models of disease [27, 28], the use of antagonist peptides incorporates elements of unpredictability that may significantly undermine their usefulness *in vivo*. Various studies have, for instance, revealed the exquisite specificity of this class of APL for individual T cell clones [3, 29]: what may be considered an effective antagonist of one TCR may, therefore, serve as a potent agonist of another. Consequently, reliance on antagonists for the treatment of disease risks igniting as many unwanted immune responses as these agents promise to extinguish. Furthermore, Gebe and co-workers [30] have demonstrated that APL of a diabetogenic epitope from GAD65 are incapable of antagonizing all the T cell clones derived from a single diabetic individual. More importantly, those T cells that proved to be refractory consistently sabotaged antagonism among susceptible clones through their secretion of IL-2, thereby ensuring that the responsive state predominates.

While such findings are poor predictors of success in a clinical context, the use of *partial* agonists may prove more efficacious. Alterations in the cytokine profile of autoreactive T cells upon recognition of analogues of their cognate peptide *in vitro* have been reported in various experimental systems [31, 32], inspiring their systemic use *in vivo* to induce immune deviation in the response to human collagen type IV [33] and to modulate the clinical course of experimental autoimmune encephalomyelitis (EAE) in rodents. In one such study, Nicholson and colleagues [34] showed how treatment of SJL mice with an APL of the immunodominant epitope of proteolipid protein ($PLP_{139-151}$) rendered them resistant to the induction of disease upon immunization with the wild-type peptide due to changes in the cytokine profile of responding cells. Significantly, adoptive transfer of Th2 cell lines, generated by priming with the APL, protected against EAE, strongly suggesting that creation of a protective cytokine milieu is able to hold in check rare T cell clones that perceive the APL as a potent agonist.

It was this underlying rationale that pervaded the first Phase II clinical trials of APL of MBP_{83-99} for the treatment of multiple sclerosis. Although both trials were prematurely curtailed due to hypersensitivity reactions among a proportion of patients [35, 36], some signs of clinical benefit were observed in the form of a reduced number and volume of enhancing lesions within the CNS, which appeared to correlate with a predominantly Th2 type response to the APL [36]. Interestingly, follow-up studies of this cohort of patients revealed persistent evidence of immune deviation up to 4.5 years after initial treatment [37]: whether or not the protracted influence of the APL will yield sustained clinical benefits in the future remains to be seen.

While immune deviation from a pathogenic Th1 response towards an anti-inflammatory cytokine profile carries a compelling logic, recent advances in our understanding of the way in which the immune system polices its own activity have suggested that alternative avenues may prove even more productive. The discovery of regulatory T cells (Treg) that actively suppress erroneous immune responses has not only illuminated a critical mechanism of self tolerance but has provided ample opportunities for harnessing their properties for therapeutic purposes, a number of strategies for immune intervention having been shown to bolster the regulatory repertoire [38]. One such property is the propensity Treg display for so-called linked suppression, the ability to influence responses to additional antigen specificities not present in the original tolerizing regime, but subsequently presented by the same antigen-presenting cell (APC) [39]. Should it prove possible to design APL capable of polarizing naïve T cells towards a Treg phenotype, linked suppression may help to broaden the scope of the tolerant state to counter the tendency for determinant spreading during the course of the autoimmune response.

That such a goal may ultimately prove feasible has been suggested by various lines of evidence that have converged on the conclusion that naïve $CD4^+$ T cells may be persuaded to adopt a regulatory phenotype following chronic, incomplete signaling through the TCR [38, 40]. Since partial agonists are, by definition, incapable of delivering full activation signals, their prolonged administration to recipients may be predicted to solicit a regulatory response. A recent study showing how predominant signaling through the ERK pathway favors not only Th2 development but also polarization towards a $CD4^+CD25^+Foxp3^+$ regulatory phenotype [41] is clearly consistent with such a prediction.

Myasthenia gravis (MG) and its experimental counterpart in mice (EAMG) have provided an ideal model in which to study the propensity of APL to exploit this natural form of self tolerance (reviewed in [42]). A few years ago, Mozes and co-workers [43] designed a 'dual APL' based on the sequences of two myasthenogenic epitopes of the nicotinic acetylcholine receptor (AChR) represented by residues 195-212 and 259-271 and linked covalently in tandem. Using a T cell line specific for $AChR_{195-212}$, this dual APL was found to inhibit proliferation and secretion of the cytokines IL-2 and IFN-γ in response to the wild-type epitope, provoking secretion of TGF-β1 and IL-10 instead. Importantly, the resulting $CD4^+CD25^+$ cells up-regulated CTLA-4 and Foxp3, indicative of their polarization towards a regulatory phenotype, and were shown to behave accordingly in co-cultures with T cells that had not been exposed to the APL. These findings have since been extrapolated to the amelioration of EAMG in susceptible strains of mice, the beneficial effects of the dual APL correlating with the appearance of Treg in the periphery [44]. Most significant, however, was the finding that the same APL was capable of modulating the responses of autoreactive T cells from patients with MG. Of 22 patients whose PBL responded when challenged with the native AChR, 21 showed significant inhibition following prior exposure to the APL, their secretion of IFN-γ being actively replaced by TGF-β1 [45], suggesting the dual APL to be a likely future candidate for immunotherapy.

Although much of the benefit of Treg cells lies in their capacity for linked suppression, formal evidence of the ability of APL to tap into this property is currently lacking. Nevertheless, there are encouraging signs that such a strategy may ultimately prove feasible: working with the dual APL, Venkata Aruna and colleagues [46] demonstrated its capacity to abrogate the onset of EAMG following immunization with the native AChR, suggesting that responses to unrelated, subdominant epitopes of the autoantigen were equally inhibited. In a similar way, Nicholson et al. [47] used an APL of $PLP_{139-151}$ to protect mice from the induction of EAE, not only upon immunization with the wild-type epitope, but also with the unrelated myelin antigens MBP and myelin oligodendrocyte glycoprotein (MOG). While these findings offer much hope for the future use of APL in a clinical context, the challenges of autoimmunity should never be underestimated: in particular, the entrenched nature of the immune response at the time of presentation, together with the likelihood of determinant spreading, poses significant obstacles. Although the immune response associated with allograft rejection is not short of complexity, the precision with which the immunological challenge may be predicted and the opportunity to condition the recipient immune system in advance of transplantation, offers hope for the use of APL in this highly specialized context.

Immune intervention in alloreactivity

Transplantation is currently the treatment of choice for end-stage organ failure, the rejection of tissues being partially mitigated by matching of donor and recipient at selected MHC loci so as to reduce the overall burden of alloreactivity. Nevertheless, residual disparities remain, including the plethora of so-called minor histocompatibility (mH) antigens whose recognition as foreign contributes to the process of chronic rejection. In recent years, many of these mH antigens have begun to yield their identity, many being shown to be peptides derived from retroviral, mitochondrial or polymorphic proteins presented to the T cell repertoire in the conventional manner by recipient MHC molecules [48]. The HA-1 antigen has, for instance, been identified as a nonamer peptide derived from the product of the KIAA0223 gene, which exists in two allelic forms differing at a single amino acid residue: the variant incorporating His instead of Arg as one of its anchor residues binds HLA-B60 with high affinity, sensitizing T cells from individuals expressing HLA-B60 to $HA\text{-}1^+$ tissues [49]. While the outcome of such T cell recognition is managed clinically by the judicious use of immune suppression, elucidation of the amino acid sequences of critical mH antigens has raised the possibility of designing APL capable of sabotaging the alloimmune response (reviewed in [50]).

This principle has been amply demonstrated in a number of *in vitro* models of alloreactivity, the first proof-of-concept deriving from an unexpected source. Working in the Hb system, Daniel and co-workers [51] made the serendipitous finding that a T cell clone specific for $Hb^d_{64\text{–}76}$ in the context of $H\text{-}2E^k$ was also alloreactive, cross-reacting with $H\text{-}2E^p$ and its cargo of endogenous peptides. Interestingly, APL of Hb, known to antagonize the T cell response to the wild-type epitope, also potently inhibited the alloresponse to $H\text{-}2E^p$ when both ligands were presented by the same APC. These findings have since been extended to the human: APL were, for instance, based on the primary sequence of the immunodominant epitope of HLA-DR β1*0101, presented in the context of HLA-DR β1*1101. Two distinct analogues were designed that bound to soluble DR11 but failed to stimulate a DR1-specific T cell clone, antagonizing, instead, the response to the native epitope

[52]. Several studies have successfully extrapolated these findings to MHC class I-restricted epitopes, significantly inhibiting cytolysis by alloreactive T cell clones [53, 54]. Perhaps most importantly, however, den Haan et al. [55] identified several APL of the HLA-A2-restricted epitope of HA-1 and demonstrated their ability to inhibit lysis of an HA-1-expressing cell line by three unrelated CTL clones. Furthermore, these APL were able to antagonize HA-1-specific polyclonal CTL lines from three patients and actively decrease the number of IFN-γ-secreting CTL in a patient with ongoing graft *versus* host disease.

Whereas the success of these studies could be largely attributed to a detailed knowledge of the immunodominant epitopes of the alloantigens on which the APL were based, such a level of insight is something of a luxury in most cases of transplantation within the human population. Nevertheless, elegant experiments by de Koster and colleagues [56] demonstrated the feasibility of designing effective antagonists of alloreactive T cell clones without prior knowledge of their cognate peptide. Using a synthetic peptide library, this group was able to identify surrogate ligands for the clones which were potently stimulatory: using first principles, these ligands were subsequently modified at critical TCR contact sites to yield APL capable of successfully antagonizing the original T cell clones.

Despite these encouraging signs, all such studies have focused exclusively on the rational design of TCR antagonists, whose recognized limitations in the autoimmunity arena are equally likely to apply to the field of transplantation. In contrast, the potential for deploying partial agonists in the fight against allograft rejection has been little explored to date, although our own studies of the immune response to the male-specific mH antigen, Dby, have highlighted their likely potential. We made use of a novel TCR-transgenic strain, the A1.RAG1$^{-/-}$ mouse, in which all T cells are specific for the immunodominant epitope of Dby (479-493) presented by H-2E^k [57]. Consequently, female A1.RAG1$^{-/-}$ mice recognize male tissues as foreign and robustly reject skin grafts with a mean survival time of 12 days. We generated APL incorporating conservative changes in the secondary TCR contact sites, and identified a number of partial agonists. One such APL, bearing His instead of Arg at residue 490, proved to be only weakly immunogenic when presented to naïve A1 T cells: nevertheless, when administered to female mice around the time of grafting, this partial agonist secured the indefinite survival of male skin grafts due to its capacity to polarize antigen-specific T cells towards a Treg phenotype [58]. Although the propensity for linked suppression could not be formally addressed in such a TCR-transgenic model, the form of tolerance observed displayed the classical features of regulation, including resistance to infusion by a second cohort of naïve female A1 T cells. These results strongly suggest that by ensuring persistent, incomplete signaling through the TCR, APL based on the sequences of epitopes from known alloantigens may be used to establish a network of Treg that may be effectively harnessed in the service of transplantation tolerance.

Conclusions

During the 15 years since their first description, APL have provided an effective means of infiltrating the immunological synapse. This subtle form of intervention has been responsible for illuminating the way in which T cells perceive their ligand and challenging our out-dated models of T cell activation. By providing a way of influencing down-stream signaling cascades and cell fate decisions, the potential offered by APL for intervening in erroneous immune responses has attracted considerable interest. Although the road ahead may be long and arduous, the ability of partial agonists to recruit Treg cells with their unique capacity for linked suppression, offers hope for specific immune intervention in autoimmunity and allograft rejection.

Acknowledgements
I am grateful to colleagues, past and present, for all their helpful discussions relating to the potential use of APL, but especially to David Wraith, with whom I worked on autoimmunity and to Herman Waldmann and Stephen Cobbold, for their insights into allograft rejection.

References

1 Fairchild PJ, Wraith DC (1992) Peptide-MHC interactions in autoimmunity. *Curr Opin Immunol* 4: 748–753
2 De Magistris MT, Alexander J, Coggeshall M, Altman A, Gaeta FCA, Grey HM, Sette A (1992) Antigen-analog-major histocompatibility complexes act as antagonists of the T cell receptor. *Cell* 68: 625–634
3 Jameson SC, Carbone FR, Bevan MJ (1993) Clone-specific T cell receptor antagonists of MHC class I-restricted cytotoxic T lymphocytes. *J Exp Med* 177: 1541–1550
4 Alexander J, Snoke K, Ruppert J, Sidney J, Wall M, Southwood S, Oseroff C, Arrhenius T, Gaeta FCA, Colon SM et al (1993) Functional consequences of engagement of the T cell receptor by low affinity ligands. *J Immunol* 150: 1–7
5 Sloan-Lancaster J, Allen PM (1996) Altered peptide ligand-induced partial T cell activation: molecular mechanisms and role in T cell biology. *Annu Rev Immunol* 14: 1–27
6 Evavold BD, Sloan-Lancaster J, Allen PM (1993) Tickling the T cell receptor: selective T cell functions stimulated by altered peptide ligands. *Immunol Today* 14: 602–609
7 Evavold BD, Allen PM (1991) Separation of IL-4 production from Th cell proliferation by an altered T cell receptor ligand. *Science* 252: 1308–1310
8 Evavold BD, Sloan-Lancaster J, Hsu BL, Allen PM (1993) Separation of T helper 1 clone cytolysis from proliferation and lymphokine production using analog peptides. *J Immunol* 150: 3131–3140

9 Reis e Sousa C, Levine EH, Germain RN (1996) Partial signalling by CD8+ T cells in response to antagonist ligands. *J Exp Med* 184: 149–157

10 Sloan-Lancaster J, Evavold BD, Allen PM (1993) Induction of T cell anergy by altered T cell receptor ligand on live antigen presenting cells. *Nature* 363: 156–159

11 Sloan-Lancaster J, Evavold BD, Allen PM (1994) Th2 cell clonal anergy as a consequence of partial activation. *J Exp Med* 180: 1195–1205

12 Tao X, Grant C, Constant S, Bottomly K (1997) Induction of IL-4 producing CD4+ T cells by antigenic peptides altered for T cell receptor binding. *J Immunol* 158: 4237–4244

13 Fairchild PJ (1997) Altered peptide ligands: prospects for immune intervention in autoimmune disease. *Eur J Immunogenet* 24: 155–167

14 Lyons DS, Lieberman SA, Hampl J, Boniface JJ, Chien Y, Berg LJ, Davis MM (1996) A T cell receptor binds to antagonist ligands with lower affinities and faster dissociation rates than to agonists. *Immunity* 5: 53–61

15 Sykulev Y, Vugmeyster Y, Brunmark A, Ploegh HL, Eisen HN (1998) Peptide antagonism and T cell receptor interactions with peptide-MHC complexes. *Immunity* 9: 475–483

16 La Face DM, Couture C, Anderson K, Shih G, Alexander J, Sette A, Mustelin T, Altman A, Grey HM (1997) Differential T cell signaling induced by antagonist peptide-MHC complexes and the associated phenotypic responses. *J Immunol* 158: 2057–2064

17 Smyth LA, Ardouin L, Williams O, Norton T, Tybulewicz V, Kioussis D (2002) Inefficient clustering of tyrosine-phosphorylated proteins at the immunological synapse in response to an antagonist peptide. *Eur J Immunol* 32: 3386–3394

18 Ehrlich LIR, Ebert PJR, Krummel MF, Weiss A, Davis MM (2002) Dynamics of p56lck translocation of the T cell immunological synapse following agonist and antagonist stimulation. *Immunity* 17: 809–822

19 Huang J, Tilly D, Altman A, Sugie K, Grey HM (2000) T cell receptor antagonists induce Vav phosphorylation by selective activation of Fyn kinase. *Proc Natl Acad Sci USA* 97: 10923–10929

20 Ding Y-H, Baker BM, Garboczi DN, Biddison WE, Wiley DC (1999) Four A6–TCR/peptide/HLA-A2 structures that generate very different T cell signals are nearly identical. *Immunity* 11: 45–56

21 Robertson JM, Evavold BD (1999) Dueling T cell receptors: peptide antagonism of CD4+ T cells with dual antigen specificities. *J Immunol* 163: 1750–1754

22 Daniels MA, Schober SL, Hogquist KA, Jameson S C (1999) A test of the dominant negative signal model for T cell receptor antagonism. *J Immunol* 162: 3761–3764

23 Yang W, Grey HM (2003) Study of the mechanism of T cell receptor antagonism using dual T cell receptor-expressing T cells. *J Immunol* 170: 4532–4538

24 Sloan-Lancaster J, Shaw AS, Rothbard JB, Allen PM (1994) Partial T cell signaling: altered phosphor-ζ and lack of ZAP-70 recruitment in APL-induced T cell anergy. *Cell* 79: 913–922

25 Madrenas J, Wange RL, Wang JL, Isakov N, Samelson LE, Germain RN (1995) ζ phos-

phorylation without ZAP-70 activation induced by T cell receptor antagonists or partial agonists. *Science* 267: 515–518
26 Singh RAK, Zhang JZ (2004) Differential activation of ERK, p38, and JNK required for Th1 and Th2 deviation in myelin-reactive T cells induced by altered peptide ligand. *J Immunol* 173: 7299–7307
27 Franco A, Southwood S, Arrhenius T, Kuchroo VK, Grey HM, Sette A, Ishioka GY (1994) T cell receptor antagonist peptides are highly effective inhibitors of experimental allergic encephalomyelitis. *Eur J Immunol* 24: 940–946
28 Karin N, Mitchell DJ, Brocke S, Ling N, Steinman L (1994) Reversal of EAE by a soluble peptide variant of a myelin basic protein epitope: T cell receptor antagonism and reduction of interferon γ and TNFα production. *J Exp Med* 180: 2227–2237
29 Ostrov D, Krieger J, Sidney J, Sette A, Concannon P (1993) T cell receptor antagonism mediated by interaction between TCR junctional residues and peptide antigen analogues. *J Immunol* 150: 4277–4283
30 Gebe JA, Masewicz SA, Kochik SA, Reijonen H, Nepom GT (2004) Inhibition of altered peptide ligand-mediated antagonism of human GAD65-responsive $CD4^+$ T cells by non-antagonizable T cells. *Eur J Immunol* 34: 3337–3345
31 Windhagen A, Scholz C, Höllsberg P, Fukaura H, Sette A, Hafler DA (1995) Modulation of cytokine patterns of human autoreactive T cell clones by a single amino acid substitution of their peptide ligand. *Immunity* 2: 373–380
32 Nicholson LB, Waldner H, Carrizosa AM, Sette A, Collins M, Kuchroo VK (1998) Heteroclitic proliferative responses and changes in cytokine profile induced by altered peptides: implications for autoimmunity. *Proc Natl Acad Sci USA* 95: 264–269
33 Pfeiffer C, Stein J, Southwood S, Ketelaar H, Sette A, Bottomly K (1995) Altered peptide ligands can control CD4 T lymphocyte differentiation *in vivo*. *J Exp Med* 181: 1569–1574
34 Nicholson LB, Greer JM, Sobel RA, Lees MB, Kuchroo VK (1995) An altered peptide ligand mediates immune deviation and prevents autoimmune encephalomyelitis. *Immunity* 3: 397–405
35 Bielekova B, Goodwin B, Richert N, Cortese I, Kondo T, Afshar G, Gran B, Eaton J, Antel J, Frank JA et al (2000) Encephalitogenic potential of the myelin basic protein peptide (amino acids 83-99) in multiple sclerosis: Results of a phase II clinical trial with an altered peptide ligand. *Nat Med* 6: 1167–1175
36 Kappos L, Comi G, Panitch H, Oger J, Antel J, Conlon P, Steinman L (2000) Induction of a non-encephalitogenic type 2 T helper-cell autoimmune response in multiple sclerosis after administration of an altered peptide ligand in a placebo-controlled, randomized phase II trial. *Nat Med* 6: 1176–1182
37 Kim HJ, Antel JP, Duquette P, Alleva DG, Conlon PJ, Bar-Or A (2002) Resistance of immune responses to altered and native myelin antigens in patients with multiple sclerosis treated with altered peptide ligand. *Clin Immunol* 104: 105–114
38 Waldmann H, Chen T-C, Graca L, Adams E, Daley S, Cobbold S, Fairchild PJ (2006) Regulatory T cells in transplantation. *Semin Immunol* 18: 111–119

39 Davies JD, Leong LY, Mellor A, Cobbold SP, Waldmann, H (1996) T cell suppression in transplantation tolerance through linked suppression. *J Immunol* 156: 3602–3607

40 Chen T-C, Cobbold SP, Fairchild PJ, Waldmann H (2004) Generation of anergic and regulatory T cells following prolonged exposure to a harmless antigen. *J Immunol* 172: 5900–5907

41 Ben-David H, Venkata Aruna B, Sela M, Mozes E (2007) A dual altered peptide ligand inhibit myasthenia gravis associated responses by inducing phosphorylated extracellular-regulated kinase 1,2 that upregulates $CD4^{+}CD25^{+}Foxp3^{+}$ cells. Scand *J Immunol* 65: 567–576

42 Sela M, Mozes E (2004) Therapeutic vaccines in autoimmunity. *Proc Natl Acad Sci USA* 101: 14586–14592

43 Venkata Aruna B, Sela M, Mozes E (2005) Suppression of myasthenogenic responses of a T cell line by a dual altered peptide ligand by induction of $CD4^{+}CD25^{+}$ regulatory cells. *Proc Natl Acad Sci USA* 102: 10285–10290

44 Ben-David H, Sela M, Mozes E (2005) Down-regulation of myasthenogenic T cell responses by a dual altered peptide ligand *via* $CD4^{+}CD25^{+}$-regulated events leading to apoptosis. *Proc Natl Acad Sci USA* 102: 2028–2033

45 Dayan M, Sthoeger Z, Neiman A, Abarbanel J, Sela M, Mozes E (2004) Immunomodulation by a dual altered peptide ligand of autoreactive responses to the acetylcholine receptor of peripheral blood lymphocytes of patients with myasthenia gravis. *Hum Immunol* 65: 571–577

46 Venkata Aruna B, Sela M, Mozes E (2006) Down-regulation of T cell responses to AChR and reversal of EAMG manifestations in mice by a dual altered peptide ligand *via* induction of $CD4^{+}CD25^{+}$ regulatory cells. *J Neuroimmunol* 177: 63–75

47 Nicholson LB, Murtaza A, Hafler BP, Sette A, Kuchroo VK (1997) A T cell receptor antagonist peptide induces T cells that mediate bystander suppression and prevent autoimmune encephalomyelitis induced with multiple myelin antigens. *Proc Natl Acad Sci USA* 94: 9279–9284

48 Roopenian D, Choi EY, Brown A (2002) The immunogenomics of minor histocompatibility antigens. *Immunol Rev* 190: 86–94

49 den Haan J, Meadows LM, Wang W, Pool J, Blokland E, Bishop TL, Reinhardus C, Shabanowitz J, Offringa R, Hunt DF et al (1998) The minor histocompatibility antigen HA-1: A diallelic gene with a single amino acid polymorphism. *Science* 279: 1054–1057

50 Fairchild PJ, Chen T-C, Waldmann H (2004) Altered peptide ligands as therapeutic agents for the treatment of allograft rejection. *Curr Top Pept Protein Res* 6: 237–244

51 Daniel C, Grakoui A, Allen PM (1998) Inhibition of an *in vitro* $CD4^{+}$ T cell alloresponse using altered peptide ligands. *J Immunol* 160: 3244–3250

52 Colovai AI, Liu Z, Harris PE, Cortesini R, Suciu-Foca N (1997) Allopeptide-specific T cell reactivity altered by peptide analogs. *J Immunol* 158: 48–54

53 Burrows SR, Khanna R, Moss DJ (1999) Direct alloreactivity by human CTL can be inhibited by altered peptide ligand antagonism. *Blood* 93: 1020–1024

54 García-Peydro M, Paradela A, Albar JP, López de Castro JA (2000) Antagonism of

direct alloreactivity of an HLA-B27–specific CTL clone by altered peptide ligands of its natural epitope. *J Immunol* 165: 5680–5685

55 den Haan JMM, Mutis T, Blokland E, Ijzerman AP, Goulmy E (2002) General T cell receptor antagonist to immunomodulate HLA-A2-restricted minor histocompatibility antigen HA-1-specific T cell responses. *Blood* 99: 985–992

56 de Koster HS, Vermeulen CJ, Hiemstra HS, Amons R, Drijfhout JW, Koning F (1999) Definition of agonists and design of antagonists for alloreactive T cell clones using synthetic peptide libraries. *Int Immunol* 11: 585–591

57 Zelenika D, Adams E, Mellor A, Simpson E, Chandler P, Stockinger B, Waldmann H, Cobbold SP (1996) Rejection of H-Y disparate skin grafts by monospecific $CD4^+$ Th1 and Th2 cells: no requirement for $CD8^+$ T cells or B cells. *J Immunol* 161: 1868–1874

58 Chen T-C, Waldmann H, Fairchild PJ (2004) Induction of dominant transplantation tolerance by an altered peptide ligand of the male antigen Dby. *J Clin Invest* 113: 1754–1762

Targeting CD4 for the induction of dominant tolerance

Herman Waldmann, Elizabeth Adams and Stephen Cobbold

Sir William Dunn School of Pathology, Oxford University, South Parks Road, Oxford OX1 3RE, UK

The immunogenicity of therapeutic antibodies as the starting point for investigating the tolerogenic potential of CD4 antibodies

It was long known from the work of Chiller and Weigle, in the 1970s [1], that deaggregated (monomeric) foreign immunoglobulins were poorly immunogenic, and capable of inducing tolerance. Aggregated immunoglobulins were, in contrast, very immunogenic.

The early 1980 saw a growing interest in developing rodent monoclonal antibodies as human therapeutics. Although monomeric these were "foreign" proteins, and one could not have predicted whether they would behave as tolerogens or as immunogens. We tested the immunogenicity of a number of monomeric rat antibodies injected into mice [2]. Those that were unable to bind to mouse blood cells behaved as tolerogens, while all those that bound proved immunogenic. The exception were rat CD4 antibodies that appeared immunologically silent, a finding also made by David Wofsy [3].

This "silence" was, it emerged, a result of CD4 antibodies inducing tolerance to themselves. We examined whether this tolerizing property of CD4 antibodies could be generalized, and found that they could induce tolerance to other aggregated "foreign" immunoglobulins, and indeed to other therapeutic antibodies [2, 4]. Using CD4 antibody F(ab')$_2$ fragments [5], or non-lytic CD4 antibodies [6, 7], it became clear that that T helper cell depletion was not essential to tolerance. This ruled out our inital hypothesis that we were simply reducing T helper cell numbers and the opportunity for T cells to form collaborative units [8].

This state of tolerance was fully dependent on peripheral mechanisms, as it could be induced in adult-thymectomized mice [7]. It came as some surprise that tolerance could not be broken by infusion of naive spleen cells, unless host CD4 T cells were depleted in advance [9]. We termed this state one of 'resistance'. Resistance implied that some CD4-dependent regulatory mechanism was contributing to tolerance. The propensity for tolerance was not restricted to the naive host, but could also be induced in a mouse previously "primed" to low doses of the foreign antigen [9]. (Higher antigen doses generated antibody responses that obscured the readout, and shortened the half-life of the proposed tolerogen.) This capacity to induce tolerance

The Immune Synapse as a Novel Target for Therapy, edited by Luis Graca

to a foreign protein was, it turned out, not unique to CD4, and could later be demonstrated with LFA-1 (CD11a) [8] and CD40L (CD154) antibodies [10].

We realized that foreign antigens would, eventually, be cleared from the body. If so, we would expect tolerance to lapse over time. Indeed, that was shown to be the case, unless the animals were regularly re-exposed to the foreign protein every few weeks, in which case tolerance could be sustained indefinitely [7]. Repeated exposure to otherwise immunogenic forms of antigen was shown to maintain or reinforce the tolerant state.

Healthy tissues are reliable sources of persisting antigens

The finding that the antigen was needed to maintain tolerance led us to wonder what would happen to transplanted tissues placed under the CD4 antibody umbrella? After all, once accepted, they would be able to provide the necessary source of antigen to maintain tolerance. Our earlier studies with depleting CD4 and CD8 antibodies had led us to conclude that both subsets participated in the rejection process [11], and that their depletion could give long-term skin-graft survival well beyond the period that the therapeutic antibodies remained in the body.

We combined CD4 and CD8 antibodies as a short-term umbrella to smuggle in a foreign bone marrow (B10.BR marrow into CBA/Ca mice) that differed antigenically across multiple minor transplantation antigens [12]. As it turned out, donor chimerism (albeit at a low level) was achieved, and this was associated with transplantation tolerance to subsequent administration of donor skin. Animals were also compromised in their capacity to reject skin bearing the foreign "minors" associated with host-type MHC. This suggested that the host might also have tolerance to "indirectly" presented (or processed) donor alloantigens. We observed, once again, that T helper cell depletion was not essential, that "resistance" was present, and that a proportion of T cells with specificity for donor antigens were still present in the host, yet unresponsive (anergic) to donor antigens *in vitro* [12]. This finding of anergic CD4 T cells combined with "resistance" led us to the hypothesis that the anergic $CD4^+$ T cells might be regulators of the immune system, as "civil servants" that would occupy critical sites of antigen presentation and prevent useful collaboration [13]. This was, of course, well before the finding that $CD4^+CD25^+$ regulatory T cells (Treg) behaved as if anergic *in vitro*! [14, 15].

T cell-mediated regulation as a mechanism in therapeutic CD4 antibody-induced tolerance

So, by 1989, we were anticipating a regulatory mechanism as part explanation of tolerance. In retrospect, we were somewhat fortunate that the marrow dose we had chosen had only given such a low level of donor chimerism, as we might have missed that evidence of regulation. Our later and more detailed analysis demonstrated that the same antibody protocol could enable a high level of donor chimerism if the marrow input was large enough [16], but that the state of tolerance achieved was "deletional" with minimal regulation detectable. At the other extreme, very low doses of marrow could prime for regulation, yet without the need for any sustained chimerism [16].

It soon emerged that the combination of non-lytic CD4 and CD8 antibodies (co-receptor blockade) could be used to tolerize mice to the tissue grafts themselves, without any need for the intermediate marrow inoculum [7]. Tolerance could be induced in adult thymectomized mice, so emphasizing its peripheral nature, and again, was associated with resistance. T cell transfer studies demonstrated that host T cells took some weeks to behave as if tolerant [17], and that the failure to reject early on must have resulted from the ceasefire brought about by the coreceptor antibodies blocking $CD4^+$ and $CD8^+$ T cell function. After this time the transferred T cells behaved as if tolerant, but persistence of that tolerant state did require re-exposure to antigen [17–19]. Without antigen, tolerance was eventually lost. Once tolerance had been established host CD4 T cells could be shown to be suppressive to naive and primed T cells (either $CD4^+$ or $CD8^+$) in adoptive transfer studies [20, 21].

Infectious tolerance and linked suppression

Using genetically tagged T cells we were able to show that the capacity to mediate resistance was a property of the tolerized host T cells [22, 23]. Ablation of the tagged cells eliminated resistance. Co-existence of naive T cells with host T cells within the tolerant graft-bearing host led to the donor T cells becoming tolerant in their own right, and exhibiting resistance in their own right [22]. We coined this state, one of "infectious tolerance". This predicted that there was a mechanism by which some naive T cells might be continuously converted to regulatory function in circumstances where the graft persisted. In other words, regulation might not be maintained by the first cohort of regulatory T cells that were generated, but by new cohorts generated throughout the life of the graft.

One of the criteria that we use to determine transplantation tolerance is that the host (A) bears the donor (B) without the need for any maintenance immunosuppression, while remaining competent to reject a third party graft (C). One of the features of tolerance mediated through coreceptor blockade is that the tolerant state can extend to third party antigens if those antigens are expressed within the grafts bearing tolerated antigens [24]. We coined this phenomenon "linked suppression". To pin down how third party antigens co-expressed with tolerated ones might operate to exploit regulation, we examined the contributions of donor antigens that had been reprocessed by host dendritic cells (DC) (indirect presentation). Experiments using genetically defined donor and recipient sets led us to conclude that "linked-suppression" could operate by host T cells recognizing donor antigens reprocessed on host-type DC [25]. In short, if both donor and third-party antigens were in the same graft they would likely be reprocessed by the same host DC within the graft. This would not be the case for the two grafts (B and C) placed in the same graft bed, where the only prospect for co-localization would be within the draining lymph nodes.

One interpretation for linked suppression was that the CD4$^+$ regulatory T cells were somehow drawn to the graft, and only there could they impact T cells recognizing third party (C) antigens. An ingenious experiment performed by Luis Graca [26] tested that idea. He transferred tolerated grafts to animals whose T cells had been depleted, and whose capacity to regenerate their T cells had been compromised. After allowing graft-derived T cells to recolonize the lymphopenic host, one could then examine whether any resistance had been acquired. This could be assessed by infusing a fresh cohort of naive T cells and providing a new donor skin graft. The outcome of this study was the conclusion that tolerated grafts contained T cells that could regulate, and that these were capable of recolonizing the peripheral pool. This was the first recognition that Treg colonized tolerated grafts, and was later confirmed directly by analysis of graft-infiltrating cells for the growing band of Treg-associated markers (FoxP3, GITR, CD25) [27–29], and attempts made to exploit the finding for diagnostic purposes aimed at providing an indication of long-term graft outcome. Much later we were able to provide evidence (by ablative studies) that the graft Treg were contributing to a privileged environment within the graft-such that T cells with the potential to reject were simply unable to deliver the destructive cargoes [30].

A reductionist approach to mechanism

The availability of TCR transgenic mice where all CD4 T cells bear an identical specificity for the target antigen, and additionally of FoxP3 as a marker for Treg, gave us the opportunity to re-examine tolerance processes following co-receptor blockade [31].

As predicted, tolerance was not accompanied by any numerical loss of CD4 T cells. It was, however, associated with evidence of resistance and of induction (conversion) of naïve T cells to cells expressing FoxP3, and other Treg-associated markers [29]. Strikingly, the frequency of such cells was far higher in the tolerated grafts than in the spleen of tolerant recipients. Tolerated grafts not only expressed abundant FoxP3 message, but also message for mast cell genes, at much greater abundance than syngeneic control grafts [28]. This finding prompted others to examine and confirm a role for mast cells in certain forms of transplantation tolerance [32].

We attempted to determine how conversion of naive T cells to T cells expressing Foxp3 (Treg) came about in anti-CD4-treated animals. The addition of non-depleting CD4 antibody and antigenic peptide to cultures of splenocytes from our TCR transgenic mice resulted in a marked conversion of T cells to those expressing FoxP3 message [29]. On the basis of a then recent report that naive T cells incubated with agonist anti-CD3 antibody and TGF-β could convert to FoxP3^{+} T cells [33], we added anti-TGF-β to the cultures and observed that conversion had been prevented. We went back to assess whether anti-CD4-mediated tolerance could be induced if TGF-β were neutralized *in vivo*, and found that it could not. Recent studies of Stephen Daley and Jianbo Ma [39] indicate that TGF-β signals must be acting, at least in part, by signaling through T cells, as mice carrying a dominant negative receptor of TGF-β, could not be tolerized by co-receptor blockade.

Antigen as the driver for infectious tolerance

The same TCR transgenic mice discussed above have enabled us to study events that might be operating in the maintenance phase during which infectious tolerance is operating. We have reasoned that the healed graft would be releasing (constitutively) donor antigens that might be processed (indirect presentation) by host DC.

These DC would not be in any way activated, and would be presenting donor antigens constitutively in the same way as they might be presenting "self". We have recently shown that immature DC carrying donor antigens can indeed tolerize in the TCR transgenic model above, and that this form of tolerance is also associated with conversion of naive T cells to Treg [34]. This leads us to conclude that the healed and accepted graft provides a constant source of donor antigen for indirect processing, which can not only maintain Treg survival, but can also allow further Treg recruitment and conversion. This finding effectively closes the loop on how tolerance is maintained long-term after just a short period of CD4 antibody treatment.

Is CD4 antibody treatment special in respect of tolerance induction?

As indicated earlier there are many antibodies that seem able to induce dominant tolerance in a variety of different contexts. These include CD4, CD3 [35], CD11a, CD40L [36] amongst the most effective. Although CD4 antibodies may have some special features related to the generation of TGF-β, we suspect that the common feature of all tolerizing-blockading antibodies is their ability to create a ceasefire between the immune system and the targeted tissue. If during that ceasefire inflammation stops and healing begins, then antigen presentation will operate outside of an inflammatory context, and will consequently be tolerance-biased. The availability of TGF-β at sites of healing and at sites where apoptotic cells may have been taken up may enhance the rate at which naive T cells become Treg. The combination of hyporesponsiveness from danger-free antigen combined with induction/conversion of Treg, may tip the balance of immune activity such that T cells with the potential to damage, are overwhelmed, and consequently restrained. A major issue is to what extent those restraining influences operate outside the target tissues and to what extent they have to act within the tissue. If they act within the tissue, do they help that tissue acquire privilege by interacting with other hemopoietic cells (such as mast cells), and /or by cross-talk with the tissue elements themselves [37, 38]?

Therapeutic prospects

In the context of clinical application, "co-receptor blockade" has not yet achieved the celebrity status of "co-stimulation blockade", even though the detailed cellular mechanisms are better understood. In part, this may relate to the selective targeting of CD4 T cells to create the initial ceasefire, where for many diseases other lymphocyte populations may be contributing or providing sniper activity.

More likely, we suspect, the inital forays into CD4 therapy in the 1990s were all conducted without adequate knowledge of mechanism, and dosing requirements. The use of depleting antibodies, rodent rather than humanized, inadequate dosing, and other pharma-prone inadequacies may have meant their low prioritization in the scheme of things.

With a better understanding of the mechanism, we would argue that CD4 antibody therapy does have a future, especially in conjunction with rational choices of synergistic agents that promote rather than inhibit the critical tolerance processes. This therapy should benefit many diseases from allergies to a broad range of autoimmune conditions in the first instance, and perhaps transplantation in a properly constructed combination.

References

1. Chiller JM, Weigle WO (1971) Cellular events during induction of immunologic unresponsiveness in adult mice. *J Immunol* 106: 1647–1653
2. Benjamin RJ, Cobbold SP, Clark MR, Waldmann H (1986) Tolerance to rat monoclonal antibodies. Implications for serotherapy. *J Exp Med* 163: 1539–1552
3. Gutstein NL, Seaman WE, Scott JH, Wofsy D (1986) Induction of immune tolerance by administration of monoclonal antibody to L3T4. *J Immunol* 137: 1127–1132
4. Benjamin RJ, Waldmann H (1986) Induction of tolerance by monoclonal antibody therapy. *Nature* 320: 449–451
5. Carteron NL, Wofsy D, Seaman WE (1988) Induction of immune tolerance during administration of monoclonal antibody to L3T4 does not depend on depletion of $L3T4^+$ cells. *J Immunol* 140: 713–716
6. Qin S, Cobbold S, Tighe H, Benjamin R, Waldmann H (1987) CD4 monoclonal

antibody pairs for immunosuppression and tolerance induction. *Eur J Immunol* 17: 1159–1165

7 Qin SX, Wise M, Cobbold SP, Leong L, Kong YC, Parnes JR, Waldmann H (1990) Induction of tolerance in peripheral T cells with monoclonal antibodies. *Eur J Immunol* 20: 2737–2745

8 Benjamin RJ, Qin SX, Wise MP, Cobbold SP, Waldmann H (1988) Mechanisms of monoclonal antibody-facilitated tolerance induction: a possible role for the CD4 (L3T4) and CD11a (LFA-1) molecules in self-non-self discrimination. *Eur J Immunol* 18: 1079–1088

9 Wise M, Benjamin R, Qin S, Cobbold S, Waldmann H (1992) Tolerance induction in the peripheral immune system. In: H Vogel, F Alt (eds): *Molecular Mechanisms of Immunological Self-Recognition*, Academic Press, New York, 149–155

10 Honey K, Cobbold SP, Waldmann H (1999) Dominant regulation: a common mechanism of monoclonal antibody induced tolerance? *Immunol Res* 20: 1–14

11 Cobbold S, Waldmann H (1986) Skin allograft rejection by L3/T4$^+$ and Lyt-2$^+$ T cell subsets. *Transplantation* 41: 634–639

12 Qin SX, Cobbold S, Benjamin R, Waldmann H (1989) Induction of classical transplantation tolerance in the adult. *J Exp Med* 169: 779–794

13 Waldmann H, Qin S, Cobbold S (1992) Monoclonal antibodies as agents to reinduce tolerance in autoimmunity. *J Autoimmun* 5 (Suppl A): 93–102

14 Sakaguchi S, Sakaguchi N, Asano M, Itoh M, Toda M (1995) Immunologic self-tolerance maintained by activated T cells expressing IL-2 receptor alpha-chains (CD25). Breakdown of a single mechanism of self-tolerance causes various autoimmune diseases. *J Immunol* 155: 1151–1164

15 Thornton AM, Shevach EM (1998) CD4$^+$CD25$^+$ immunoregulatory T cells suppress polyclonal T cell activation *in vitro* by inhibiting interleukin 2 production. *J Exp Med* 188: 287–296

16 Bemelman F, Honey K, Adams E, Cobbold S, Waldmann H (1998) Bone marrow transplantation induces either clonal deletion or infectious tolerance depending on the dose. *J Immunol* 160: 2645–2648

17 Scully R, Qin S, Cobbold S, Waldmann H (1994) Mechanisms in CD4 antibody-mediated transplantation tolerance: kinetics of induction, antigen dependency and role of regulatory T cells. *Eur J Immunol* 24: 2383–2392

18 Cobbold SP, Qin S, Leong LY, Martin G, Waldmann H (1992) Reprogramming the immune system for peripheral tolerance with CD4 and CD8 monoclonal antibodies. *Immunol Rev* 129: 165–201

19 Cobbold SP, Adams E, Marshall SE, Davies JD, Waldmann H (1996) Mechanisms of peripheral tolerance and suppression induced by monoclonal antibodies to CD4 and CD8. *Immunol Rev* 149: 5–33

20 Marshall SE, Cobbold SP, Davies JD, Martin GM, Phillips JM, Waldmann H (1996) Tolerance and suppression in a primed immune system. *Transplantation* 62: 1614–1621

21 Davies JD, Martin G, Phillips J, Marshall SE, Cobbold SP, Waldmann H (1996) T cell regulation in adult transplantation tolerance. *J Immunol* 157: 529–533

22 Qin S, Cobbold SP, Pope H, Elliott J, Kioussis D, Davies J, Waldmann H (1993) "Infectious" transplantation tolerance. *Science* 259: 974–977
23 Chen ZK, Cobbold SP, Waldmann H, Metcalfe S (1996) Amplification of natural regulatory immune mechanisms for transplantation tolerance. *Transplantation* 62: 1200–1206
24 Davies JD, Leong LY, Mellor A, Cobbold SP, Waldmann H (1996) T cell suppression in transplantation tolerance through linked recognition. *J Immunol* 156: 3602–3607
25 Wise MP, Bemelman F, Cobbold SP, Waldmann H (1998) Linked suppression of skin graft rejection can operate through indirect recognition. *J Immunol* 161: 5813–5816
26 Graca L, Cobbold SP, Waldmann H (2002) Identification of regulatory T cells in tolerated allografts. *J Exp Med* 195: 1641–1646
27 Cobbold SP, Adams E, Graca L, Waldmann H (2003) Serial analysis of gene expression provides new insights into regulatory T cells. *Semin Immunol* 15: 209–214
28 Cobbold SP, Nolan KF, Graca L, Castejon R, Le Moine A, Frewin MR, Humm S, Adams E, Thompson S, Zelenika D et al (2003) Regulatory T cells and dendritic cells in transplantation tolerance: molecular markers and mechanisms. *Immunol Rev* 196: 109–124
29 Cobbold SP, Castejon R, Adams E, Zelenika D, Graca L, Humm S, Waldmann H (2004) Induction of foxP3$^+$ regulatory T cells in the periphery of T cell receptor transgenic mice tolerized to transplants. *J Immunol* 172: 6003–6010
30 Cobbold SP, Adams E, Graca L, Daley S, Yates S, Paterson A, Robertson NJ, Nolan KF, Fairchild PJ, Waldmann H (2006) Immune privilege induced by regulatory T cells in transplantation tolerance. *Immunol Rev* 213: 239–255
31 Zelenika D, Adams E, Humm S, Lin CY, Waldmann H, Cobbold SP (2001) The role of CD4$^+$ T cell subsets in determining transplantation rejection or tolerance. *Immunol Rev* 182: 164–179
32 Lu LF, Lind EF, Gondek DC, Bennett KA, Gleeson MW, Pino-Lagos K, Scott ZA, Coyle AJ, Reed JL, Van Snick J et al (2006) Mast cells are essential intermediaries in regulatory T cell tolerance. *Nature* 442: 997–1002
33 Chen W, Wahl SM (2003) TGF-beta: the missing link in CD4($^+$)CD25($^+$) regulatory T cell-mediated immunosuppression. *Cytokine Growth Factor Rev* 14: 85–89
34 Yates SF, Paterson AM, Nolan KF, Cobbold SP, Saunders NJ, Waldmann H, Fairchild PJ (2007) Induction of regulatory T cells and dominant tolerance by dendritic cells incapable of full activation. *J Immunol* 179: 967–976
35 Chatenoud L (2003) CD3-specific antibody-induced active tolerance: from bench to bedside. *Nat Rev Immunol* 3: 123–132
36 Honey K, Cobbold SP, Waldmann H (1999) CD40 ligand blockade induces CD4$^+$ T cell tolerance and linked suppression. *J Immunol* 163: 4805–4810
37 Waldmann H, Adams E, Fairchild P, Cobbold S (2006) Infectious tolerance and the long-term acceptance of transplanted tissue. *Immunol Rev* 212: 301–313
38 Waldmann H (2006) Immunology: protection and privilege. *Nature* 442: 987–988
39 Daley SR, Ma J, Adams E, Cobbold SP, Waldmann H (2007) A key role for TGF{beta} signalling to T-cells in the long-term acceptance of allografts. *J Immunol* 179; *in press*

Anti-CD3: from T cell depletion to tolerance induction

Damien Bresson and Matthias von Herrath

La Jolla Institute for Allergy and Immunology, Department of Developmental Immunology 3, 9420 Athena Circle, La Jolla, CA 92037, USA

Immunosuppressive drugs: The ice age of immune interventions

Immune processes need to be strictly controlled to counteract any immunological disorders or pathological events and maintain a healthy balance in the body. Consequently, a variety of immune interventions have been preclinically validated and revealed great promise in animal models. However, the translation from bench to bedside has been more than disappointing in various clinical trials. The first therapeutic agents (immune suppressors) were mostly nonspecific and inhibited cellular proliferation [1–3]. These treatments generally led to serious side effects due intrinsic lack of pharmacospecificity. Later, cyclosporin A (CsA) was the first of a new generation of immunosuppressants with a 'site-specific' mode of action. Mechanistically, CsA mediates its *in vivo* effect by repressing lymphocyte activation at an early stage. Due to a low degree of myelotoxicity, CsA was considered early as an attractive therapeutic drug in clinical transplantation for inhibiting lymphocytic activities without affecting either phagocytosis or migration of the reticulo-endothelial system. In 1978, CsA was tested clinically and due to its strong efficacy was readily used worldwide in a majority of the transplant centers to maintain graft survival post surgery [4, 5]. In the mean time, much work has been put into the design of new therapeutic strategies that would present lower side effects but retain substantial efficacy.

The birth of anti-CD3 antibodies

A major advance in the development of selective immune-targeting drugs was the development in 1975 of a method enabling the generation of monoclonal antibodies (mAbs) *in vitro* [6]. For the first time it was possible to target specifically several antigens and move towards safer immune interventions. In spite of the production of a plethora of different mAbs directed against various surface or intracellular antigens, few of them underwent clinical evaluations due to a lack of efficacy or

The Immune Synapse as a Novel Target for Therapy, edited by Luis Graca

some adverse side effects that would have been unacceptable in humans. Among these mAbs, CD3-specific antibodies were found particularly successful in delaying or treating several immune disorders (Tab. 1). The first murine anti-CD3 mAb was produced by the Schlossman group in 1979 and was called OKT3 [7]. OKT3 is a mouse IgG2a subtype antibody binding specifically to the epsilon subchain of the CD3 complex expressed by both $CD4^+$ and $CD8^+$ T cells [8, 9]. The CD3 molecule functions as a molecular bridge holding the T cell receptor (TCR) to the cell surface. This enables the signal delivered through the TCR and peptide/major histocompatibility complex (MHC) at the surface of antigen-presenting cells (APC), and leading to T cell activation events. Recently, two teams were able to co-crystallize and solve the structure of the human CD3$\varepsilon\gamma$ in complex with an antigen-binding fragment (Fab) of OKT3 [8, 9]. Inspection of the crystal structure revealed that OKT3 mAb interacts with a conformational epitope located on the CD3epsilon subunit. Mechanistically, OKT3 as well as other anti-CD3-specific mAbs induce a strong mitogenic response promoting a general T cell expansion and cytokine production both *in vitro* and *in vivo* [10–13].

Clinically, in treated patients, administration of OKT3 caused a 'flu-like' syndrome associated with transient symptoms such as fever, headache, nausea, vomiting and gastrointestinal disturbance. Those are triggered by the binding of the crystallized fragment (Fc) portion of the mAb to Fc receptor-bearing cells provoking a strong systemic release of both Th1 (TNF-α, IFN-γ, IL-2) and Th2 (IL-6 and IL-10) cytokines [14–20]. Other side effects, described in rodents or in humans, are inherent to strong immunosuppression and comprise (i) partial lymphopenia (as long as the antibody is found in the body), (ii) virus infection or reactivation (mainly cytomegalovirus and Epstein-Barr virus), but also (iii) production of human anti-mouse antibodies (HAMA) [21–26]. It is worth mentioning that preclinical studies were generally performed with a hamster anti-mouse CD3 mAb (namely 145-2C11) [27].

Similarly to that observed with OKT3, 145-2C11 mAb binds to the epsilon chain of the CD3 complex and interacts with all $CD3^+$ T cells with an ability to modulate T cell functions.

OKT3 mediates short-term T cell depletion

In 1981, Russell and colleagues [25] reported the first clinical data using OKT3 in renal transplantation. In this Phase I/II clinical trial in which eight patients received cadaver renal transplant, kidney allograft rejection was ameliorated by a short-term treatment (1–5 mg intravenous infusion for 10–20 days) with OKT3 mAb in conjunction with immunosuppressors (azathioprine and prednisone). Kidney transplant rejection was stopped by a rapid $CD3^+$ T cell depletion accompanied by relatively minor side effects including chills and fever. Unfortunately, this 'grace' period was not long lasting and during the next 3- to 12-month follow-up period further rejec-

Table 1 - Examples of anti-CD3 antibodies tested in clinical trials

Year	Antibody name	Antibody species	Trial	Clinical indication	Refs.
1981	OKT3	Mouse IgG_{2A} anti-CD3	Phase I/II	Acute renal allograft rejection	[25]
1982			Phase II	GVHD	[26]
1984			Phase II	Acute renal allograft rejection	[28]
1985			Phase II	Acute renal allograft rejection	[24]
1987			Phase II	Acute renal allograft rejection	[30]
1991			Phase II	MS	[22]
1999	hOKT3γ1 (Ala-Ala)	Mutated humanized IgG_1	Phase I	Kidney transplant	[34]
2004			Phase I	Islet transplantation	[36]
2002			Phase II	Psoriatic arthritis	[35]
2002			Phase II/III	T1DM	[59]
1997 and 2002	Visilizumab (HuM291)	Mutated humanized IgG_2	Phase I	Bone marrow transplantation	[39, 40]
2000			Phase I	Kidney transplant	[72]
2003	T3/4.A	Mouse IgA	Phase II	Kidney transplant	[41]
1999	Campath 3 (YTH12.5 or ChAglyCD3)	Aglycosylated humanized IgG1	Phase I	Kidney transplant	[37]
2005			Phase II/III	T1DM	[66]

Ig, Immunogobulin; GVHD, Graft-versus-host disease; MS, multiple sclerosis; T1DM, type 1 diabetes mellitus.

tions occurred in a majority of patients. Later, another clinical trial was conducted to study the efficacy of OKT3.PAN (T3) mAb in acute cadaveric renal allograft rejection [28]. Acute rejection was reversed in eight out of nine patients by a 14-day treatment with T3 mAb (5 mg/day administered intravenously). In accordance with the positive outcomes observed in these first clinical trials, in 1984 the U.S. Food and Drug Agency and other regulatory authorities worldwide approved the use of OKT3 (commercialized under the name of Orthoklone and manufactured by Ortho-Biotech) for treating acute kidney transplant rejection.

Although, several trials were conducted in the U.S. or in Europe with an indisputable success [24, 29–33], the related side effects (such as a cytokine storm syndrome) often described upon OKT3 treatment weakened its therapeutic value in humans.

Engineered anti-CD3 antibodies: preserving efficacy while avoiding strong side effects

To circumvent systemic side effects, anti-CD3 mAbs were engineered to avoid interaction with the Fc receptors, i.e., CD16, CD32 and CD64. A series of non-Fc binding anti-CD3 was developed and tested clinically in kidney, islet and bone marrow transplantation (Tab. 1). They all induce transient lymphodepletion lasting from a couple of days to a couple of weeks after treatment has ended.

Humanized IgG_1 anti-CD3

A humanized IgG_1 form of OKT3 [hOKT3γ1(Ala-Ala)] was mutated in the Fc domain where amino acid residues 234 and 235 were replaced by alanines to avoid binding to Fc receptors [34]. This mAb has been used in various clinical trials. For instance, in the field of transplantation seven patients were treated daily with hOKT3γ1(Ala-Ala) (5–10 mg/day) for 10 consecutive days to achieve serum levels of 1 μg/ml. Among them, five patients showed a rapid reversal of rejection, which was prolonged over a year without strong side effects. Later, in psoriatic arthritis (PsA), a chronic disease characterized by inflammation of the skin (psoriasis) and joints (arthritis), the efficacy of hOKT3γ1(Ala-Ala) mAb was evaluated in a Phase I/II clinical trial [35]. Seven patients were treated with increasing daily doses of anti-CD3 mAb for 12–14 consecutive days. A short-term decrease of the symptoms (such as inflamed joints and pain scale) was described in six out of seven patients. Unfortunately, at day 90 after treatment only two out of six responders had sustained improvement. No patients developed strong side effects; however, at the highest hOKT3γ1(Ala-Ala) concentration, mild cytokine release symptoms associated with elevation of IL-10 were detected. A forthcoming Phase II clinical trial will establish the *bona fide* efficacy and safety of the drug in patients suffering from PsA.

In the field of autoimmune diabetes, one of the most attractive avenues to reverse hyperglycemia is transplantation of insulin-secreting beta-cells into diabetic patients. Although, islet transplantation protocols have been improved, clinicians are still seeking for immunomodulating agents that could prolong graft survival with low adverse events. To reach this goal, the hOKT3γ1(Ala-Ala) mAb was applied for 12 consecutive days (4 mg/day), beginning 2 days before their islet allograft transplants [36]. Four out of six patients achieved and maintained insulin independence with normal metabolic control.

Aglycosylated humanized anti-CD3 IgG_1

Nine patients received a short-term treatment with an aglycosylated humanized anti-CD3 IgG_1 (campath 3 or ChAglyCD3) for 8 consecutive days at 8 mg/day [37].

None of the patients demonstrated any anti-globulin response or any significant cytokine release syndrome. Almost 78% showed proof of resolution of their rejection, although some patients experienced re-rejection.

Humanized anti-CD3 IgG_2

A Phase I dose escalation was performed using a humanized IgG_2 anti-CD3 (HuM291 or visilizumab), engineered to lower any mitogenic activity in humans [38, 39]. A single dose of 0.015 mg/kg was well tolerated with only mild to moderated side effects and was sufficient to induce T cell depletion for up to 1 week post-treatment. Later, in a Phase II clinical trial, the HuM291 mAb showed great promise for the treatment of acute graft-*versus*-host disease (GVHD); GVHD is mediated by donor T cells and presents a major barrier to successful hematopoietic cell transplant. The risk-benefit ratio was found to be acceptable with a single-dose regiment of HuM291 at a dosage ranging from 0.5 to 6.15 mg/patient. The drug was well tolerated and some signs of GVHD amelioration were observed in a majority of recipients [40]. Further trials need to be performed to determine the efficacy of visilizumab in GVHD.

Murine anti-CD3 IgA antibody

A non-mitogenic murine IgA antibody binding to human CD3 (T3/4.A) was tested in a Phase II clinical trial [41]. Fifteen patients were enrolled and received an intravenous injection with 5 mg /day for 10 consecutive days. Most of the patients developed transient vomiting and/or diarrhea, which coincided with elevated serum levels of proinflammatory cytokines. These side effects disappeared after antibody clearance.

Anti-CD3 antibodies as mediators of self tolerance in autoimmune disorders

Type 1 diabetes

Transplantation and autoimmunity share a number of important immunological pathways, explaining the ability of allograft rejection to trigger autoimmune responses or the increased susceptibility of patients with autoimmune diseases to allograft rejection. Therefore, with regard to the potent effect of anti-CD3 mAbs in the field of transplantation, it was highly relevant to evaluate their effect in autoimmunity. In the early nineties, the group of Hayward and Shreiber observed for the first time that a single neonatal injection with anti-CD3 mAb induces immune tolerance by modulating the T cell repertoire and stopping/delaying autoimmunity in non-obese diabetic (NOD) mice, genetically predisposed to type 1 diabetes (T1D) [42]. From this date, much work has been accomplished in unraveling the mechanisms involved in such a therapeutic potency.

During pathogenesis of T1D, autoreactive $CD4^+$ and $CD8^+$ T cells are generated and progressively destroy the insulin-producing pancreatic beta-cells. The destruction of approximately 80% of beta-cells has to occur both in animal models and in humans before T1D becomes symptomatic. In the past decades, a series of immune therapies have been elaborated to treat T1D with some encouraging results. However, development of a cure for T1D is particularly difficult, because it mostly affects young adults and children, therefore the ethical window for any treatment is rather small and long-term side effects have to be avoided. Furthermore, insulin injected into the body as a palliative therapy affords a reasonable life quality and expectancy. However, insulin cannot prevent all of the late complications of T1D, and the life expectancy can be reduced by 10–15 years due to serious clinical complications [43]. Thus, production of non-mitogenic anti-CD3 mAbs that are deprived of strong side effects, resurrected the interest of the scientists in these molecules.

First, the anti-CD3 mAb (clone 145-2C11) was engineered as a non-Fc binding $F(ab')_2$ for preclinical studies. Short-course treatment with this mAb was shown to reverse T1D in hyperglycemic NOD mice [44, 45]. Therapeutic efficacy was related to two striking features. First, the treatment was most efficient when administered into already diabetic animals. This was highly unusual since more than 200 treatments were capable of preventing T1D but very few can reverse it after hyperglycemia has occurred [46]. Second, in contrast to that observed with strong immunosuppressive agents, long-term immune suppression was not needed to maintain permanent tolerance to beta-cell autoantigens (aAgs). A 5-day course of therapy after onset with low dose anti-CD3 $F(ab')_2$ was sufficient to cure diabetes in a majority of mice and hyperglycemia did not recur over time. It is worth noting that efficacy of anti-CD3 was not mouse strain dependent since a similar protection was reported in the transgenic rat insulin promoter-lymphocytic choriomeningitis virus (RIP-LCMV) mice, a second model where T1D is induced upon infection with LCMV [47]. In subsequent studies, the group of J. F. Bach and L. Chatenoud sheds light on potential mechanisms involved in the anti-diabetogenic effect observed with anti-CD3 $F(ab')_2$. They demonstrated that it induced active tolerance mediated by regulatory T cells (Tregs) expressing the surface markers CD4 (co-receptor in the immune synapse), CD25 (IL-2 receptor) and CD62L (lymphocyte adhesion molecule 1: L-selectin). When co-transferred with diabetogenic effector T cells into immunocompromised NOD-SCID mice, these Tregs protected from diabetes [44, 45, 48–56]. Such a potent protective effect upon adoptive transfer was not observed with immunocompetent RIP-LCMV mice, which is a more severe model for T1D [47, 57]. These observations raise the paramount question of whether a systemic immune modulator such as anti-CD3 acting on virtually all T cells, and not only on islet-specific T cells, can expand a sufficient number of islet-specific Tregs *in vivo* to induce full protection when transferred into immunocompetent recipients. However, it does emphasize the fact that the tolerogenic capacity of anti-CD3-specific mAb involves two phases to be fully functional [49, 50]. The first induction phase, lasting approximately for a week after ending antibody injection, is associated with a direct action on effector T cells. The insulitis in anti-CD3-treated mice is rapidly cleared within 2 or 3 days, leading to normoglycemia. Then, a second phase involving an expansion of Tregs is mandatory to maintain permanent tolerance to beta-cells aAgs. Therefore, adoptive transfer of anti-CD3-induced Tregs into immunocompetent mice only mimics the second phase of the treatment and does not reflect the full protective capacity of anti-CD3 therapy. Mechanistically, the transforming growth factor-β (TGF-β) secreted by the anti-CD3 expanded Tregs, but not IL-4, plays a central role in the restoration of peripheral active tolerance [58].

In 2000, the groups of J. A. Bluestone and K. C. Herold initiated a clinical trial in patients suffering from recent-onset T1D [59]. A total of 24 patients were enrolled in an open-label control trial and randomized to each of the study group: A 14-day course treatment with the hOKT3γ1(Ala-Ala) mAb or placebo control. All patients underwent a mixed meal-tolerance test and other immunological studies every 6 months. Thanks to the mutations in the Fc region of the mAb, the adverse events that occurred with drug administration were generally mild and included most commonly, rash, fever and other 'flu-like' symptoms but of less severity than those following administration of OKT3. Therapeutically, over a 24-month period, a single course treatment within the first 6 weeks after diagnosis significantly preserved the C-peptide response, a cleavage product from the processing of proinsulin to insulin measured to differentiate insulin produced by the body from insulin injected into the body as a palliative therapy [60]. Improvement in the C-peptide levels was also accompanied by amelioration in glucose control reflected by HbA1c level as well as lower exogenous insulin requirements. At a cellular level, hOKT3γ1(Ala-Ala) therapy significantly augmented IL-10 and IL-5 cytokines in the peripheral blood of responsive patients while IFN-γ and IL-6 cytokine levels were decreased [61, 62]. Phenotypic studies of peripheral lymphocytes revealed a higher number of IL-10 expressing $CD4^+$ T cells after anti-CD3 treatment. These cells were heterogeneous but generally $CD45RO^+$ (a memory marker), $CD25^+$, and $CD62L^-$, and expressed CCR4 (CC chemokine receptor 4). More surprisingly, suppressor $CD8^+CD25^+$ Tregs were identified in clinical responders and expanded after therapy [63–65]. These cells were CTLA-4^+ (cytotoxic T lymphocyte-associated antigen-4, encoding a receptor involved in the control of T cell proliferation and apoptosis) and $Foxp3^+$ (Forkhead box P3, a transcription-repressor protein) and required cell-cell contact for inhibition.

In light of the success obtain with the hOKT3γ1(Ala-Ala) mAb, a European multicenter trial was conducted with the aglycosylated ChyAglyCD3 mAb (Tab. 1). Two major conclusions can be drawn from the first report published 18 months after treatment [66]. First, short-term therapy with ChyAglyCD3 mAb preserved residual beta-cell function in patients with new-onset T1D and showing the highest beta-cell mass at trial entry (C-peptide levels superior to the 50th percentile). Second, the adverse side effects observed in the European trial were more severe that the ones reported in the American trial. Administration with ChAglyCD3 was associated with moderate 'flu-like' symptoms and transient but generalized Epstein-Barr viral reactivation. Such activation of latent virus particles was probably due to an increase in the anti-CD3 dose, from 28 to 48 mg/patient in the American and European trials, respectively, which should be considered in future clinical applications with any non-Fc binding anti-CD3 mAbs.

Multiple sclerosis

In 2005, the group of S. D. Miller extended the therapeutic efficacy of non-mitogenic anti-CD3 using an experimental autoimmune encephalomyelitis (EAE) animal model for human multiple sclerosis (MS). Similarly to that described for T1D, when injected intravenously, the Fc-altered anti-CD3 mAbs reversed new-onset EAE [67]. However, two striking differences with the diabetes settings were found. First, although protection correlated with an increase in the frequency of $CD4^+CD25^+$ T cells neither anti-CD25 nor anti-TGF-β antibody treatment abrogated the efficacy. Second, as recently reported, protection was mediated by $CD4^+CD25^-$ T cells expressing the latency-associated peptide (LAP) on their surface, thus confirming a negligible role for $CD4^+CD25^+$ Tregs [68].

Conclusions and perspectives

In less than 30 years, anti-CD3 therapy has rapidly imposed itself in the world of immune interventions. These mAbs have been mainly used to prolong graft survival or prevent/treat various immune syndromes. Broad clinical use of anti-CD3 therapy was made possible thanks to the generation of non-Fc binding antibodies, which lowered the adverse side effects of treatment. Their mode of action possesses two main features that distinguish them from conventional immunosuppressive agents. First, upon short-course treatment a rapid lymphodepletion of auto-reactive $CD3^+$ T cell is observed and lasts for a couple of weeks after ending treatment. In a second phase, self tolerance is restored by resetting the immune system *via* an expansion of adaptive $CD4^+$ Tregs. Despite major advances in the prevention of acute rejection in transplantation and in the treatment of some autoimmune diseases, a single course of anti-CD3 mAb does not induce permanent tolerance. Therefore, future immunointerventions using anti-CD3 mAb will aim at prolonging the efficacy without increasing the side effects. Accordingly, multiple anti-CD3 injections or combination therapies with other drugs are now envisioned to strengthen the efficacy of the treatment. For instance, in the field of autoimmune diabetes, short-term therapy with anti-CD3 145-2C11 mAb in combination with a proinsulin vaccine reversed T1D more forcefully than the mono-therapies alone in two animal models [57, 69]. In addition, anti-CD3 can be administered together with drugs, such as exendin-4, that can exacerbate beta-cell regeneration [70, 71]. To conclude, if administered safely anti-CD3 therapy might greatly improve the management and treatment of several immunological disorders in the near future.

Acknowledgements
This work was supported by NIH grants AI51973 and DK51091 to M.G.V.H. D.B. is a recipient of a European Marie-Curie Outgoing Fellowship (2005–2008).

References

1 Monaco AP (1989) Immunosuppression and tolerance for clinical organ allografts. *Curr Opin Immunol* 1: 1174–1177
2 Collier SJ (1989) Immunosuppressive drugs. *Curr Opin Immunol* 2: 854–858
3 Fritsche L, Einecke G, Fleiner F, Dragun D, Neumayer HH, Budde K (2004) Reports of large immunosuppression trials in kidney transplantation: room for improvement. *Am J Transplant* 4: 738–743
4 Powles RL, Barrett AJ, Clink H, Kay HE, Sloane J, McElwain TJ (1978) Cyclosporin A for the treatment of graft-*versus*-host disease in man. *Lancet* 2: 1327–1331
5 Calne RY, White DJ, Thiru S, Evans DB, McMaster P, Dunn DC, Craddock GN, Pentlow BD, Rolles K (1978) Cyclosporin A in patients receiving renal allografts from cadaver donors. *Lancet* 2: 1323–1327
6 Kohler G, Milstein C (1975) Continuous cultures of fused cells secreting antibody of predefined specificity. *Nature* 256: 495–497
7 Kung P, Goldstein G, Reinherz EL, Schlossman SF (1979) Monoclonal antibodies defining distinctive human T cell surface antigens. *Science* 206: 347–349
8 Salmeron A, Sanchez-Madrid F, Ursa MA, Fresno M, Alarcon B (1991) A conformational epitope expressed upon association of CD3-epsilon with either CD3-delta or CD3-gamma is the main target for recognition by anti-CD3 monoclonal antibodies. *J Immunol* 147: 3047–3052
9 Kjer-Nielsen L, Dunstone MA, Kostenko L, Ely LK, Beddoe T, Mifsud NA, Purcell AW, Brooks AG, McCluskey J, Rossjohn J (2004) Crystal structure of the human T cell receptor CD3 epsilon gamma heterodimer complexed to the therapeutic mAb OKT3. *Proc Natl Acad Sci USA* 101: 7675–7680
10 Van Wauwe JP, De Mey JR, Goossens JG (1980) OKT3: a monoclonal anti-human T lymphocyte antibody with potent mitogenic properties. *J Immunol* 124: 2708–2713

11 Davis L, Vida R, Lipsky PE (1986) Regulation of human T lymphocyte mitogenesis by antibodies to CD3. *J Immunol* 137: 3758–3767
12 Van Wauwe J, Goossens J (1981) Mitogenic actions of Orthoclone OKT3 on human peripheral blood lymphocytes: effects of monocytes and serum components. *Int J Immunopharmacol* 3: 203–208
13 Walls EV, Borghetti AF, Benzie CR, Kay JE (1984) Early events during the activation of human lymphocytes by the mitogenic monoclonal antibody OKT3. *Cell Immunol* 89: 30–38
14 Alegre M, Depierreux M, Florquin S, Najdovski T, Vandenabeele P, Abramowicz D, Leo O, Deschodt-Lanckman M, Goldman M (1990) Acute toxicity of anti-CD3 monoclonal antibody in mice: a model for OKT3 first dose reactions. *Transplant Proc* 22: 1920–1921
15 Ferran C, Sheehan K, Dy M, Schreiber R, Merite S, Landais P, Noel LH, Grau G, Bluestone J, Bach JF et al (1990) Cytokine-related syndrome following injection of anti-CD3 monoclonal antibody: further evidence for transient *in vivo* T cell activation. *Eur J Immunol* 20: 509–515
16 Hirsch R, Gress RE, Pluznik DH, Eckhaus M, Bluestone JA (1989) Effects of *in vivo* administration of anti-CD3 monoclonal antibody on T cell function in mice. II. *In vivo* activation of T cells. *J Immunol* 142: 737–743
17 Hirsch R, Gress RE, Bluestone JA (1989) Anti-CD3 antibody for autoimmune disease, a cautionary note. *Lancet* 1: 1390
18 Abramowicz D, Schandene L, Goldman M, Crusiaux A, Vereerstraeten P, De Pauw L, Wybran J, Kinnaert P, Dupont E, Toussaint C (1989) Release of tumor necrosis factor, interleukin-2, and gamma-interferon in serum after injection of OKT3 monoclonal antibody in kidney transplant recipients. *Transplantation* 47: 606–608
19 Chatenoud L, Legendre C, Ferran C, Bach JF, Kreis H (1991) Corticosteroid inhibition of the OKT3-induced cytokine-related syndrome – dosage and kinetics prerequisites. *Transplantation* 51: 334–338
20 Chatenoud L, Ferran C, Bach JF (1991) The anti-CD3-induced syndrome: a consequence of massive *in vivo* cell activation. *Curr Top Microbiol Immunol* 174: 121–134
21 Waid TH, Lucas BA, Thompson JS, Brown SA, Munch L, Prebeck RJ, Jezek D (1992) Treatment of acute cellular rejection with T10B9.1A-31 or OKT3 in renal allograft recipients. *Transplantation* 53: 80–86
22 Weinshenker BG, Bass B, Karlik S, Ebers GC, Rice GP (1991) An open trial of OKT3 in patients with multiple sclerosis. *Neurology* 41: 1047–1052
23 Sgro C (1995) Side-effects of a monoclonal antibody, muromonab CD3/orthoclone OKT3: Bibliographic review. *Toxicology* 105: 23–29
24 Ortho Multicenter Transplant Study Group (1985) A randomized clinical trial of OKT3 monoclonal antibody for acute rejection of cadaveric renal transplants. *N Engl J Med* 313: 337–342
25 Cosimi AB, Burton RC, Colvin RB, Goldstein G, Delmonico FL, LaQuaglia MP,

Tolkoff-Rubin N, Rubin RH, Herrin JT, Russell PS (1981) Treatment of acute renal allograft rejection with OKT3 monoclonal antibody. *Transplantation* 32: 535–539

26 Prentice HG, Blacklock HA, Janossy G, Bradstock KF, Skeggs D, Goldstein G, Hoffbrand AV (1982) Use of anti-T-cell monoclonal antibody OKT3 to prevent acute graft-*versus*-host disease in allogeneic bone-marrow transplantation for acute leukaemia. *Lancet* 1: 700–703

27 Leo O, Foo M, Sachs DH, Samelson LE, Bluestone JA (1987) Identification of a monoclonal antibody specific for a murine T3 polypeptide. *Proc Natl Acad Sci USA* 84: 1374–1378

28 Bowen A, Edwards LC, Gailiunas P, Helderman JH (1984) Lymphocyte function in patients treated with monoclonal anti-T3 antibody for acute cadaveric renal allograft rejection. *Transplantation* 38: 489–493

29 Goldstein G, Norman DJ, Shield CF 3rd, Kreis H, Burdick J, Flye MW, Rivolta E, Starzl T, Monaco A (1986) OKT3 monoclonal antibody reversal of acute renal allograft rejection unresponsive to conventional immunosuppressive treatments. *Prog Clin Biol Res* 224: 239–249

30 Norman DJ, Shield CF 3rd, Barry JM, Henell K, Funnell MB, Lemon J (1987) Therapeutic use of OKT3 monoclonal antibody for acute renal allograft rejection. *Nephron* 46 (Suppl 1): 41–47

31 Monaco A, Goldstein G, Barnes L (1987) Use of Orthoclone OKT3 monoclonal antibody to reverse acute renal allograft rejection unresponsive to treatment with conventional immunosuppressive regimens. *Transplant Proc* 19 (Suppl 1): 28–31

32 Canafax DM, Draxler CA (1987) Monoclonal antilymphocyte antibody (OKT3) treatment of acute renal allograft rejection. *Pharmacotherapy* 7: 121–124

33 Delmonico FL, Cosimi AB (1988) Monoclonal antibody treatment of human allograft recipients. *Surg Gynecol Obstet* 166: 89–98

34 Woodle ES, Xu D, Zivin RA, Auger J, Charette J, O'Laughlin R, Peace D, Jollife LK, Haverty T, Bluestone JA et al (1999) Phase I trial of a humanized, Fc receptor nonbinding OKT3 antibody, huOKT3gamma1(Ala-Ala) in the treatment of acute renal allograft rejection. *Transplantation* 68: 608–616

35 Utset TO, Auger JA, Peace D, Zivin RA, Xu D, Jolliffe L, Alegre ML, Bluestone JA, Clark MR (2002) Modified anti-CD3 therapy in psoriatic arthritis: a Phase I/II clinical trial. *J Rheumatol* 29: 1907–1913

36 Hering BJ, Kandaswamy R, Harmon JV, Ansite JD, Clemmings SM, Sakai T, Paraskevas S, Eckman PM, Sageshima J, Nakano M et al (2004) Transplantation of cultured islets from two-layer preserved pancreases in type 1 diabetes with anti-CD3 antibody. *Am J Transplant* 4: 390–401

37 Friend PJ, Hale G, Chatenoud L, Rebello P, Bradley J, Thiru S, Phillips JM, Waldmann H (1999) Phase I study of an engineered aglycosylated humanized CD3 antibody in renal transplant rejection. *Transplantation* 68: 1632–1637

38 Cole MS, Stellrecht KE, Shi JD, Homola M, Hsu DH, Anasetti C, Vasquez M, Tso JY

(1999) HuM291, a humanized anti-CD3 antibody, is immunosuppressive to T cells while exhibiting reduced mitogenicity *in vitro*. *Transplantation* 68: 563–571

39 Cole MS, Anasetti C, Tso JY (1997) Human IgG2 variants of chimeric anti-CD3 are nonmitogenic to T cells. *J Immunol* 159: 3613–3621

40 Carpenter PA, Appelbaum FR, Corey L, Deeg HJ, Doney K, Gooley T, Krueger J, Martin P, Pavlovic S, Sanders J et al (2002) A humanized non-FcR-binding anti-CD3 antibody, visilizumab, for treatment of steroid-refractory acute graft-*versus*-host disease. *Blood* 99: 2712–2719

41 Meijer RT, Surachno S, Yong SL, Bemelman FJ, Florquin S, Ten Berge IJ, Schellekens PT (2003) Treatment of acute kidney allograft rejection with a non-mitogenic CD3 antibody. *Clin Exp Immunol* 133: 485–492

42 Hayward AR, Shreiber M (1989) Neonatal injection of CD3 antibody into nonobese diabetic mice reduces the incidence of insulitis and diabetes. *J Immunol* 143: 1555–1559

43 Liu E, Eisenbarth GS (2002) Type 1A diabetes mellitus-associated autoimmunity. *Endocrinol Metab Clin North Am* 31: 391–410, vii-viii

44 Chatenoud L, Primo J, Bach JF (1997) CD3 antibody-induced dominant self tolerance in overtly diabetic NOD mice. *J Immunol* 158: 2947–2954

45 Chatenoud L, Thervet E, Primo J, Bach JF (1994) Anti-CD3 antibody induces long-term remission of overt autoimmunity in nonobese diabetic mice. *Proc Natl Acad Sci USA* 91: 123–127

46 Shoda LK, Young DL, Ramanujan S, Whiting CC, Atkinson MA, Bluestone JA, Eisenbarth GS, Mathis D, Rossini AA, Campbell SE et al (2005) A comprehensive review of interventions in the NOD mouse and implications for translation. *Immunity* 23: 115–126

47 von Herrath MG, Coon B, Wolfe T, Chatenoud L (2002) Nonmitogenic CD3 antibody reverses virally induced (rat insulin promoter-lymphocytic choriomeningitis virus) autoimmune diabetes without impeding viral clearance. *J Immunol* 168: 933–941

48 Chatenoud L (2002) The use of monoclonal antibodies to restore self-tolerance in established autoimmunity. *Endocrinol Metab Clin North Am* 31: 457–475, ix

49 Chatenoud L (2003) CD3 antibody treatment stimulates the functional capability of regulatory T cells. *Novartis Found Symp* 252: 279–286; discussion 286–290

50 Chatenoud L (2003) CD3–specific antibody-induced active tolerance: from bench to bedside. *Nat Rev Immunol* 3: 123–132

51 Chatenoud L (2005) CD3–specific antibodies restore self-tolerance: mechanisms and clinical applications. *Curr Opin Immunol* 17: 632–637

52 Chatenoud L, Bach JF (2005) Resetting the functional capacity of regulatory T cells: a novel immunotherapeutic strategy to promote immune tolerance. *Expert Opin Biol Ther* 5 (Suppl 1): S73–81

53 Chatenoud L, Bach JF (2005) Regulatory T cells in the control of autoimmune diabetes: the case of the NOD mouse. *Int Rev Immunol* 24: 247–267

54 Chatenoud L (2006) [Anti-CD3 monoclonal antibodies: a new step towards therapy in new-onset type 1 diabetes]. *Med Sci (Paris)* 22: 5–6

55 Chatenoud L, Salomon B, Bluestone JA (2001) Suppressor T cells – they're back and critical for regulation of autoimmunity! *Immunol Rev* 182: 149–163

56 Chatenoud L, Thervet E, Primo J, Bach JF (1992) [Remission of established disease in diabetic NOD mice induced by anti-CD3 monoclonal antibody]. *C R Acad Sci III* 315: 225–228

57 Bresson D, Togher L, Rodrigo E, Chen Y, Bluestone JA, Herold KC, von Herrath M (2006) Anti-CD3 and nasal proinsulin combination therapy enhances remission from recent-onset autoimmune diabetes by inducing Tregs. *J Clin Invest* 116: 1371–1381

58 Belghith M, Bluestone JA, Barriot S, Megret J, Bach JF, Chatenoud L (2003) TGF-beta-dependent mechanisms mediate restoration of self-tolerance induced by antibodies to CD3 in overt autoimmune diabetes. *Nat Med* 9: 1202–1208

59 Herold KC, Hagopian W, Auger JA, Poumian-Ruiz E, Taylor L, Donaldson D, Gitelman SE, Harlan DM, Xu D, Zivin RA et al (2002) Anti-CD3 monoclonal antibody in new-onset type 1 diabetes mellitus. *N Engl J Med* 346: 1692–1698

60 Herold KC, Gitelman SE, Masharani U, Hagopian W, Bisikirska B, Donaldson D, Rother K, Diamond B, Harlan DM, Bluestone JA (2005) A single course of anti-CD3 monoclonal antibody hOKT3{gamma}1(Ala-Ala) results in improvement in C-peptide responses and clinical parameters for at least 2 years after onset of type 1 diabetes. *Diabetes* 54: 1763–1769

61 Herold KC, Burton JB, Francois F, Poumian-Ruiz E, Glandt M, Bluestone JA (2003) Activation of human T cells by FcR nonbinding anti-CD3 mAb, hOKT3gamma1(Ala-Ala). *J Clin Invest* 111: 409–418

62 Herold KC (2004) Achieving antigen-specific immune regulation. *J Clin Invest* 113: 346–349

63 Bisikirska BC, Herold KC (2004) Use of anti-CD3 monoclonal antibody to induce immune regulation in type 1 diabetes. *Ann NY Acad Sci* 1037: 1–9

64 Bisikirska BC, Herold KC (2005) Regulatory T cells and type 1 diabetes. *Curr Diab Rep* 5: 104–109

65 Bisikirska B, Colgan J, Luban J, Bluestone JA, Herold KC (2005) TCR stimulation with modified anti-CD3 mAb expands $CD8^+$ T cell population and induces $CD8^+CD25^+$ Tregs. *J Clin Invest* 115: 2904–2913

66 Keymeulen B,Vandemeulebroucke E, Ziegler AG, Mathieu C, Kaufman L, Hale G, Gorus F, Goldman M, Walter M, Candon S et al (2005) Insulin needs after CD3–antibody therapy in new-onset type 1 diabetes. *N Engl J Med* 352: 2598–2608

67 Kohm AP, Williams JS, Bickford AL, McMahon JS, Chatenoud L, Bach JF, Bluestone JA, Miller SD (2005) Treatment with nonmitogenic anti-CD3 monoclonal antibody induces $CD4^+$ T cell unresponsiveness and functional reversal of established experimental autoimmune encephalomyelitis. *J Immunol* 174: 4525–4534

68 Ochi H, Abraham M, Ishikawa H, Frenkel D, Yang K, Basso AS, Wu H, Chen ML,

Gandhi R, Miller A et al (2006) Oral CD3-specific antibody suppresses autoimmune encephalomyelitis by inducing CD4(+)CD25(-)LAP(+) T cells. *Nat Med* 12: 627–635

69 Bresson D, von Herrath M (2004) Immunotherapy after recent-onset type 1 diabetes: combinatorial treatment for achieving long-term remission in humans? *Rev Diabet Stud* 1: 108–112

70 Suarez-Pinzon WL, Lakey JR, Brand SJ, Rabinovitch A (2005) Combination therapy with epidermal growth factor and gastrin induces neogenesis of human islet {beta}-cells from pancreatic duct cells and an increase in functional {beta}-cell mass. *J Clin Endocrinol Metab* 90: 3401–3409

71 Suarez-Pinzon WL, Yan Y, Power R, Brand SJ, Rabinovitch A (2005) Combination therapy with epidermal growth factor and gastrin increases {beta}-cell mass and reverses hyperglycemia in diabetic NOD mice. *Diabetes* 54: 2596–2601

72 Norman DJ, Vincenti F, de Mattos AM, Barry JM, Levitt DJ, Wedel NI, Maia M, Light SE (2000) Phase I trial of HuM291, a humanized anti-CD3 antibody, in patients receiving renal allografts from living donors. *Transplantation* 70: 1707–1712

Immune modulation by CD40L blockade

Yuan Zhai and Jerzy W. Kupiec-Weglinski

The Dumont-UCLA Transplant Center, Department of Surgery, David Geffen School of Medicine at UCLA, 10833 Le Conte Avenue, Los Angeles, CA 90095, USA

Introduction

CD40L (CD154), a member of TNF-TNFR superfamily, binds to CD40 and several integrins, and plays the key role in host immune responses [1]. Blockade of the CD40-CD40L costimulation pathway has proven to be highly effective in modulating various types of immune responses, including anti-microbial, autoimmune, alloimmune responses, allergy, as well as tissue inflammation of both antigen-specific and nonspecific types. In particular, CD154 blockade has been widely applied in organ transplant models, from rodents to primates [2]. In this review, we focus on recent literature with emphasis on *in vivo* immunological mechanisms.

Immunobiology of CD40L

CD40L is produced either as a type-II transmembrane protein of 32–33 kDa [3] or as soluble forms of 31 and 18 kDa, which retain full biological activities as the membrane-bound CD154 and can potentially act as cytokines on distal $CD40^+$ cells [4]. CD40L is a member of TNF family, which includes TNF-α, CD153, CD70, 4-1BBL, OX40L, and FasL. The gene encoding CD40L is located on the X-chromosome in both human and mouse. Its mutations have been found in clinical patients with the X-linked hyper-IgM syndrome [5, 6]. CD40L expression was originally thought to be restricted to activated T cells, mainly $CD4^+$. It has now also been identified on $CD8^+$ T cells, B cells, eosinophils, mast cells, basophils, dendritic cells, and other cell types [1, 7]. Recently, platelets were also found to express CD40L [8].

The classic receptor for CD40L is CD40, which is a 50-kDa type-I transmembrane protein [9] in the TNFR family whose members include TNFR-1 and -2, CD30, CD27, 4-1BB, OX40, and Fas. The gene encoding CD40 is located at chromosome 20 in human and chromosome 2 in mouse [10]. CD40 is expressed widely on B cells, dendritic cells, monocytes/macrophages, thymic epithelium cells, endothelial cells [11], mast cells [12], fibroblasts [13] and smooth muscle cells [14].

The Immune Synapse as a Novel Target for Therapy, edited by Luis Graca

In recent years, alternative receptors for CD40L have been identified based on the functional activities of CD40L independent of CD40. The finding in a murine asthma model that development of bronchial hyper-responsiveness was prevented by the lack of CD40L, but not by the absence of CD40, provided the first clue that CD40L might also bind to one or more other receptors [15]. In murine thrombosis studies, recombinant soluble (rs) CD40L was shown to specifically bind to purified integrin $\alpha_{IIb}\beta_3$ and to activate platelets in a β3-dependent manner. Infusion of rsCD40L restored normal thrombosis in CD40L-deficient mice, whereas rsCD40L lacking the KGD integrin-recognition sequence did not [16]. Additionally, sCD40L-induced platelet stimulation resulted in the phosphorylation of Tyr759 in the cytoplasmic domain of β3 [17]. More recently, sCD40L was also found to bind to integrin $\alpha_5\beta_1$ [18] and $\alpha_M\beta_2$ (Mac-1) [19] on human monocytes. The impact of these novel receptors of CD40L is currently limited to atherosclerosis, and has yet been explored in other types of immune responses.

Roles in adaptive immune responses

Originally identified as the molecular interaction between CD4 T helper cells and B cells, the CD40L-CD40 pathway plays a key role in regulating thymus-dependent (TD) humoral responses [20]. Patients with X-linked hyper-IgM syndrome have elevated levels of IgM; low levels of IgA, IgG, and IgE; and the absence of germinal centers; and are unable to mount TD humoral response [21]. In mice, with either antibodies (Abs) to block the interaction in WT mice or in their gene-deficient counterparts, CD40-CD40L interaction was shown to be critical in regulating B cell proliferation, Ig production, Ig class switching, rescue of B cells from apoptosis, germinal center formation, generation of B cell memory, and clonal expansion and deletion of B cells [22–24]. In addition, CD40-CD40L interaction between B and $CD4^+$ T cells can also operate the opposite way, i.e., B cells as APCs to activate naïve CD4 T cells [25].

In a resting stage, most APCs express only low levels of costimulatory molecules, such as CD80, CD86. Initial antigen encounter triggers T cell activation cascade, resulting in the up-regulation of CD40L on T cells, which ligates CD40 on APCs, leading to their full activation. This CD40L-CD40-mediated T cell-APC interaction has been shown to be critical, particularly in non-infectious situations, in rendering APCs capable of activating naive T cells, although the details remain to be fully defined [26]. Up-regulation of costimulatory molecules in APCs, e.g., CD80/86 represents one of the outcomes of this interaction. Induction of pro-inflammatory cytokines, including IL-12 in DCs, is another outcome, which is critical for the development of Th1-type immune responses [27]. However, these are not sufficient to account for all the effects of CD40L costimulation. Indeed, although exogenous IL-12 restored Th1 responses, it failed to protect CD154-deficient hosts from *Schis-*

tosoma mansoni infection [28]. As a dynamic interplay of CD40L-CD40 interaction, cross-linking of CD40L on CD4 T cells had a direct effect *in vivo*, contributing to the generation of helper function [29]. In accordance, *in vitro* ligation of CD40L on T cells significantly enhances their cytokine production, particularly, Th2 type [30]. In an autoimmune diabetes model, it was shown that CD40L-dependent priming of diabetogenic CD4$^+$ T cells could be dissociated from activation of APCs [31].

CD154-CD40 interaction provides the molecular basis for CD4 help to activate cytotoxic CD8$^+$ T cells [32, 33]. DCs have been assumed to serve as the bridge between these two cell types [34]. The molecular details of this three cell interactions have not yet been fully defined. However, it is well established that DCs need to be "licensed" by activated CD4 T cells *via* CD154-CD40 interaction to gain the capabilities of helping CD8$^+$ T cells to differentiate into cytotoxic effectors. More recently, it was demonstrated that CD4 help could also be directly delivered to CD8 T cells *via* the same CD40L-CD40 pathway using CD40-deficient APCs [35].

Roles in tissue inflammation and injury

CD40 is expressed on many non-hematopoietic cells, particularly under the pro-inflammatory conditions [36]. Cross-linking of CD40 on vascular endothelial cells activates these cells and increases expression of adhesion molecules, such as CD62E, CD106, CD54, [37, 38] as well as production of chemokines/cytokines, such as IL-6 and IL-8 [39]. This may promote extravasation and accumulation of neutrophils and other peripheral lymphocytes at the sites of inflammation. Activated T cells, as well as platelets may provide the source of CD40L for this endothelial activation. Thrombin-activated platelets are able to rapidly up-regulate CD154 expression [8, 40], and contribute to the chemotactic effects from endothelial activation/damage, independent of T cells. In a murine inflammatory bowel disease model, it was demonstrated that the initial disease inducer triggered CD40 up-regulation in the colonic vasculature and the CD40-CD40L interaction mediated the recruitment of leukocytes and platelet in the inflamed colon, promoting the intestinal inflammatory response and tissue injury [41]. The expression of CD40 on hepatocytes can be up-regulated by TNF-α and play a role in their apoptosis contributing to fulminant hepatitis [42]. This pro-apoptotic effect of CD40 activation was partially mediated by enhanced expression of FasL in hepatocytes [42–44]. CD40 activation triggers NF-κB/AP-1 signaling and induces Fas-dependent apoptosis in human intrahepatic biliary epithelial cells [43].

The development of acute ischemia reperfusion injury (IRI) in liver, brain and lung has also been shown dependent on the CD40L-CD40 interaction. Our laboratory documented in a murine liver partial warm ischemia reperfusion model that CD40L-deficient mice or WT mice treated with CD40L blockade were all protected from IRI partially due to the suppression of intrahepatic pro-inflammatory response against IR [45, 46]. Blockade of CD40L-CD40 interaction also improved post-ischemic lung injuries in an isolated rat lung perfusion model [47], which was associated with attenuated production of MIP-2 and suppression of lymphocyte activation. In a murine model of brain focal IR, platelet and leukocyte adhesion was elevated and blood/brain barrier function was compromised by middle cerebral artery occlusion in WT mice. Blood cell recruitment and increased permeability were blunted in both CD40- and CD40L-deficient mice. Infarct volume was also reduced in these deficient mice compared with WT mice [48].

CD154 blockade in organ transplantation

As CD40L-CD40 interaction plays critical role in many aspects of host immune responses, it is not surprising to observe potent immunosuppressive effects of blocking this costimulation pathway in variety of transplant models, including pancreatic islets [49], heart [50], skin [51], small bowel [52], aorta [53], bone marrow [54], cornea [55], limbs [56], and kidney (unpublished data from authors' lab). The majority of these successful cases were, however, limited to the murine models. In large animal models and human trials, much limited effects were achieved due to the reasons discussed below. Key experimental variables affecting the efficiency of CD154-targeted therapy in transplant recipients include types of allografts, donor-recipient strain combinations, recipient's immune status and adjunctive treatments.

Adjunctive therapies

Although CD154 blockade alone prolongs allograft survival, more profound therapeutic effects have been achieved by combined treatment with either donor-splenocyte transfusion (DST) or CTLA4-Ig [51, 57–59]. Adjunctive CD8 T cell blocking/depletion was utilized because of the observation that alloreactive CD8 T cells or one of their subset (e.g., asialo $GM1^{+}$) were relatively resistant to CD154 blockade [52, 60–62]. Anti-LFA-1 and anti-CD45RB were also used in combination with CD40L blockade to control alloreactive CD8 activation [63–66]. The ICOS-B7RP-1 costimulatory pathway was demonstrated recently as a requirement for the development of chronic rejection after CD40-CD154 blockade [67]. This may provide a clue to the activation mechanism of CD154 blockade-resistant CD8 T cells and aid in designing more specific adjunctive therapies for CD154 blockade. Additionally, stimulating negative T cell signal by PD-L1-Ig concurrent with blocking CD154 has been shown to facilitate allograft survival [68].

Despite side effects and lack of control of chronic vasculopathy, transplant patients are currently enjoying conventional immunosuppressive therapies, which fairly effectively control acute allograft rejection episodes. Thus, application of costimulation blockade in clinics will inevitably require the combination with conventional immunosuppressive agents. It has been widely recognized that calcineurin inhibitors may abrogate the effects of T cell costimulatory blockade [69]. However, some recent comprehensive studies have revised this view by providing new insights into this interaction [70, 71]. Cyclosporine, tacrolimus, and anti-IL-2R monoclonal antibody (mAb) therapy abrogated the effect of a single-dose protocol of anti-CD154 therapy. In contrast, rapamycin acted synergistically with anti-CD154 therapy in promoting long-term allograft survival. The addition of calcineurin inhibitors did not abolish that synergy. Intense CD154-CD40 blockade by a multiple-dose schedule of anti-CD154 resulted in long-term graft survival, profound alloreactive T cell unresponsiveness, and even overcame the opposite effects of calcineurin inhibitors. CTLA4-Ig induced long-term graft survival, and the effect was not affected by the concomitant use of any immunosuppressive drugs.

Host immunological status

As T cells are the major targets for CD40L blockade in transplant recipients, their activation status determines dependency on this particular costimulatory molecule. Accumulating data clearly indicate that only naïve, but not memory or effector, T cells require CD40L for their activation, and that not only alloantigen-primed T cells but also cross-reactive T cells primed with microbial antigens are resistant to CD154 blockade in rejecting allografts. This provides us with an explanation why CD40L blockade losses its efficacy in large animal models and humans. Indeed, outbred recipients have much higher frequency of memory T cells reactive (or cross-reactive) to alloantigens in their immune repertoire [72]. DST/CD40L treatment of naïve B10.D2 recipients induced long-term graft survival of B10.A hearts. Previous priming of donor-specific T cells through rejection of B10.A, but not third party, skin grafts prevented the effects. Moreover, adoptive transfer of $CD3^+$, $CD4^+$ or $CD8^+$ T cells from B10.A skin-graft-primed animals prevented the effects of DST/CD40L [73]. We have shown in a murine cardiac transplant model that CD40L blockade effectively prolonged allograft survival in naïve recipients, but had no effect when recipients were primed with a donor-type skin graft either 10 or 40 days earlier [74]. Depletion of CD8 T cells in these sensitized recipients reinstated the efficacy of CD40L blockade. Additionally, as T cell depletion induction regimens are often used in transplant patients, the resulting homeostatic proliferation of residual T cells leads to the generation of functional memory T cells. The latter in murine models have been found resistant to costimulatory blockade [75, 76].

The generation of cross-reactive T cells responding to alloantigens by virus infection has been known for sometime [77, 78]. However, the relevance of these activated T cells to allograft rejection and responses to costimulation blockade remains unknown. Adam et al. [79] showed that lymphocytic choriomeningitis virus (LCMV)- or vaccinia virus (VV)-induced alloantigen cross-reactive T cells were maintained as memory cells, which could respond to alloantigen stimulation with second-order kinetics. Importantly, alloreactive memory cells, either from alloantigen or viral antigen primed, prevented tolerance induction by CTLA4-Ig and anti-CD154 once beyond a certain threshold level in host peripheral repertoire in a bone marrow transplant model. Furthermore, they demonstrated that CD8 central memory T cells were the principal mediators of this costimulation blockade-resistant rejection. $CD4^+$ T cells obtained from C57BL/6 (B6) mice that clinically resolved *Leishmania major* infection exhibited statistically significant cross-reactivity toward P/J (H-2p) Ags compared with the response to other haplotypes. B6 mice that were previously infected with *L. major* specifically rejected P/J skin grafts with second-order kinetics compared with naive animals. Although DST combined with anti-CD154 Ab induced prolonged graft survival in naive animals, the same treatment was ineffective in mice previously infected with *L. major* [80].

Unlike animal models, transplant patients may experience concurrent infections, which may have significant impact on the efficacy of costimulatory blockade. Indeed, acute infection with LCMV induced allograft rejection in mice treated with DST (or sublethal irradiation) plus anti-CD154 Ab if inoculated at the time of, or shortly after, transplantation in skin or bone marrow transplant models [81, 82]. As Toll-like receptors (TLRs) are the host sentinel system responding to various infectious agents by recognizing pathogen-associated molecular pattern (PAMS), the complex interplay between host innate immune activation and alloimmune responses has been recently investigated. Thornley et al. [83] showed that TLR signaling triggered by four different TLR agonists abrogates the effects of costimulation blockade by preventing alloreactive $CD8^+$ T cell apoptosis. Similarly, Chen et al. [84] found that the engagement of a single TLR was sufficient to prevent anti-CD154-mediated long-term cardiac allograft acceptance and correlated with abolished intragraft recruitment of $CD4^+/FoxP3^+$ regulatory T cells and the development of linked suppression.

Thus, both adaptive and innate immune status influences the host alloimmune activation and its response to costimulatory blockade.

Suppression/tolerance mechanisms

Alloreactive CD8 T cells, as discussed above, were initially thought to be resistant to CD154 blockade. Indeed, disruption of CD154-CD40 signaling failed to prevent rejection of minor histocompatibility complex-mismatched skin grafts in CD4-depleted recipients, whereas nonlytical anti-CD8 in combination with CD154 blockade did induce tolerance in the same model [60]. Similar outcomes were observed in a small bowel transplant model in which anti-CD40L prolonged allograft survival only in CD8 KO, but not CD4 KO recipients [52]. The direct demonstration of CD40L-independent CD8 activation resulting in graft rejection was shown by Jones et al. [61] in a murine cardiac transplant model using CD8 TCR Tg cells. However, we have shown that CD40L blockade did target alloreactive CD8 T cells. Allogeneic skin grafts failed to induce the generation of alloreactive CD8 effectors ($CD8^{+}CD44^{high}CD62L^{low}$) in CD40L-deficient mice, while a comparable extent of $CD4^{+}$ T cell proliferation was observed *in vivo* if infused into allogeneic hosts with either WT or CD154-deficient lymphocytes. The inhibition of alloreactive CD8 T cell activation by CD40L blockade was repeated in WT B6 mice with anti-CD40L mAb [85]. In addition, we have shown that both CD4-dependent and CD4-independent activation pathways were operational for alloreactive CD8 T cells, and that both pathways were sensitive to CD40L blockade [86]. In support of our conclusion, it has been shown that hepatocyte rejection in CD4 KO mice, which is CD8 dependent, was suppressed by treatment with DST and anti-CD154 mAb therapy [87].

Which mechanisms are responsible for maintaining long-term allograft survival in recipients subjected to transient costimulation blockade? Two nonexclusive mechanisms have been documented. Deletion has been implicated in central tolerance, whereas peripheral tolerance has generally been ascribed to clonal anergy and/or active immunoregulatory states. Using mice transgenic for Bcl-xL, in which T cells were resistant to passive cell death through cytokine withdrawal, Wells et al. [88] demonstrated that CD154 blockade plus DST could not induce cardiac allograft tolerance in the absence of T cell death. Li et al. [89] also showed that blocking both CD28-B7 and CD154-CD40 interactions inhibited proliferation of alloreactive T cells *in vivo*, while allowing cell cycle-dependent T cell apoptosis of proliferating T cells, with permanent engraftment of cardiac allografts but not skin allografts. Treatment with rapamycin plus costimulation blockade resulted in massive apoptosis of alloreactive T cells and produced stable skin allograft tolerance. In contrast, treatment with cyclosporine A and costimulation blockade abolished T cell proliferation and apoptosis, as well as the induction of stable allograft tolerance. Thus, deletion of activated T cells, both CD4 and CD8, through activation-induced cell death or growth factor withdrawal seems necessary to achieve peripheral tolerance across major histocompatibility complex (MHC) barriers. Moreover, several studies documented direct depletion of activated T cells by anti-CD40L Abs *via* Fc receptor- and complement-dependent mechanisms.

While combination of DST and anti-CD154 MR1 Ab permitted permanent islet engraftment, administration of cobra venom factor (which depletes complements) abolished the tolerance induction [90]. Moreover, DST plus MR1 did not prevent acute islet allograft rejection when complement C5-deficient DBA/2 mice were used as recipients. Thus, it was concluded that the tolerizing MR1 mechanisms were not limited to the blockade of CD40/CD154 signals. The complement-dependent cytotoxicity contributed to MR1 anti-CD154-induced immunosuppression. In a murine MHC-mismatched skin transplant model, Monk et al. [91] showed that prolongation of graft survival is dependent on both complement- and Fc receptor-mediated mechanisms, suggesting that Abs to CD154 act through selective depletion of activated T cells, rather than exerting immune modulation by costimulation blockade, as currently postulated. However, whether depletion of alloreactive T cells by anti-CD40L treatment is universally necessary remains controversial. It was found that spleens of mice bearing long-term cardiac allografts following inductive anti-CD40L treatment retained precursor donor alloantigen-reactive CTL, IL-2-producing helper cells, and Th1 in numbers comparable to those observed in naive mice [92]. The combined anti-CD154/anti-LFA-1 therapy, which was highly effective for inducing long-term allograft survival in high-responder C57BL/6 recipients, did not deplete the tracer alloreactive CD8 T cells, which were TCR Tg and donor reactive [64].

Regulatory T cells (Treg cells) have been shown to play a critical role in costimulation blockade-induced allograft tolerance. First shown by Honey et al. [60], CD154 blockade plus anti-CD8 induced transplantation tolerance associated with linked suppression, a form of dominant immune regulation mediated by $CD4^+$ T cells. Taylor et al. [93, 94] further showed that $CD4^+$ T cells were tolerized to alloantigen *via ex vivo* CD154/CD40 or CD28/B7 blockade, resulting in secondary mixed leukocyte reaction hyporesponsiveness and tolerance to alloantigen *in vivo*. $CD4^+CD25^+$ T cells were found to be potent regulators of alloresponses. Depletion of $CD4^+CD25^+$ T cells from the $CD4^+$ responder population completely abrogated ex vivo tolerance induction to alloantigen as measured by intact responses to alloantigen restimulation *in vitro* and *in vivo*. Reintroduction of $CD4^+CD25^+$ T cells to $CD4^+CD25^-$ cultures restored tolerance induction [93].

We have been studying the tolerance mechanism of CD154 blockade, with particular emphasis on CD4 Treg-CD8 interaction. In our C57BL/6 recipients of BALB/c cardiac allografts, we have shown that alloreactive CD8 T cells are the major effectors [86, 95]. A single dose of anti-CD40L Ab at the time of engraftment effectively prevented the activation of alloreactive CD8 T cells and induced donor-specific immunological tolerance (long-term heart graft survival, acceptance of donor but rejection of third-party test skin graft). When test skin graft was applied to our long-term tolerant recipients, only third-party, but not donor-type, skin grafts could induce activated CD8 T cells in PBLs. Interestingly, CD4 depletion prior to the test skin graft restored donor-reactive CD8 T cell activation leading to the rejection of both donor skin and the original cardiac allograft. Similar effects were observed with anti-CD25 and anti-CTLA4 Abs. These were clearly indicative of a non-deletional, $CD4^+CD25^+$ Treg-mediated mechanism in suppressing alloreactive CD8 T cell activation in a donor-specific fashion. Similarly, by adaptive transfer of alloreactive TCR Tg $CD4^+$ T cells into skin graft recipients, Quezada et al. [96] showed that CD40L blockade with DST prevented the expansion of these T cells without depletion. At the tolerant stage, blockade of Treg function readily induced the activation of these alloreactive T cells and subsequent graft rejection. However, in CD40L blockade-induced bone marrow chimeras, the initial CD4 suppression/anergy was followed by peripheral depletion of alloreactive CD4 T cells [97].

Summary

The roles that the CD40L-CD40 costimulatory pathway are known to play in host immune responses are expanding, as more cell types are found to express these molecules, and signaling in this pathway operates in both ways. Thus, the known functions of CD40L-CD40 interaction are extended from its original adaptive T/B cell activation to the innate inflammation/tissue injury. Although its blockade has initially been proven therapeutically effective in many disease states, host conditions may drastically alter the therapy outcome. One example is the resistance of effector/memory type T cells to this blockade, which has been documented not only in transplant but also in allergy models [98]. Although innate immune activation *via* TLRs may circumvent the requirement of CD40L costimulation by T cells, the underlying mechanism remains to be determined. The controversy regarding the effect of CD40L blockade on alloreactive CD8 T cells may be resolved based on the heterogeneity of CD8 population, existence of distinctive resistant subsets or differential responses by CD8 T cells at different differentiation status, such as naïve, memory/effector (direct or cross-reactive). Both depletion and immune regulation mechanisms contribute to the long-term effects of CD154 blockade. Despite the progresses, however, more careful and comprehensive research is warranted to fully elucidate the functional mechanism of CD40L in the host immunity.

References

1 Grewal IS, Flavell RA (1998) CD40 and CD154 in cell-mediated immunity. *Annu Rev Immunol* 16: 111–135

2 Yamada A, Sayegh MH (2002) The CD154-CD40 costimulatory pathway in transplantation. *Transplantation* 73: S36–39

3 Armitage RJ, Fanslow WC, Strockbine L, Sato TA, Clifford KN, Macduff BM, Anderson DM, Gimpel SD, Davis-Smith T, Maliszewski CR et al (1992) Molecular and biological characterization of a murine ligand for CD40. *Nature* 357: 80–82

4 Ludewig B, Henn V, Schroder JM, Graf D, Kroczek RA (1996) Induction, regulation, and function of soluble TRAP (CD40 ligand) during interaction of primary $CD4^+$ $CD45RA^+$ T cells with dendritic cells. *Eur J Immunol* 26: 3137–3143

5 Villa A, Notarangelo LD, Di Santo JP, Macchi PP, Strina D, Frattini A, Lucchini F, Patrosso CM, Giliani S, Mantuano E et al (1994) Organization of the human CD40L gene: implications for molecular defects in X chromosome-linked hyper-IgM syndrome and prenatal diagnosis. *Proc Natl Acad Sci USA* 91: 2110–2114

6 Notarangelo LD, Peitsch MC (1996) CD40lbase: a database of CD40L gene mutations causing X-linked hyper-IgM syndrome. *Immunol Today* 17: 511–516
7 Mach F, Schonbeck U, Sukhova GK, Bourcier T, Bonnefoy JY, Pober JS Libby P (1997) Functional CD40 ligand is expressed on human vascular endothelial cells, smooth muscle cells, and macrophages: implications for CD40-CD40 ligand signaling in atherosclerosis. *Proc Natl Acad Sci USA* 94: 1931–1936
8 Henn V, Slupsky JR, Grafe M, Anagnostopoulos I, Forster R, Muller-Berghaus G, Kroczek RA (1998) CD40 ligand on activated platelets triggers an inflammatory reaction of endothelial cells. *Nature* 391: 591–594
9 Stamenkovic I, Clark EA, Seed B (1989) A B-lymphocyte activation molecule related to the nerve growth factor receptor and induced by cytokines in carcinomas. *EMBO J* 8: 1403–1410
10 Grimaldi JC, Torres R, Kozak CA, Chang R, Clark EA, Howard M, Cockayne DA (1992) Genomic structure and chromosomal mapping of the murine CD40 gene. *J Immunol* 149: 3921–3926
11 Yellin MJ, Brett J, Baum D, Matsushima A, Szabolcs M, Stern D, Chess L (1995) Functional interactions of T cells with endothelial cells: the role of CD40L-CD40-mediated signals. *J Exp Med* 182: 1857–1864
12 Krishnaswamy G, Lakshman T, Miller AR, Srikanth S, Hall K, Huang SK, Suttles J, Smith JK, Stout R (1997) Multifunctional cytokine expression by human mast cells: regulation by T cell membrane contact and glucocorticoids. *J Interferon Cytokine Res* 17: 167–176
13 Yellin MJ, Winikoff S, Fortune SM, Baum D, Crow MK, Lederman S, Chess L (1995) Ligation of CD40 on fibroblasts induces CD54 (ICAM-1) and CD106 (VCAM-1) upregulation and IL-6 production and proliferation. *J Leukoc Biol* 58: 209–216
14 Lazaar AL, Amrani Y, Hsu J, Panettieri RA Jr, Fanslow WC, Albelda SM, Pure E (1998) CD40–mediated signal transduction in human airway smooth muscle. *J Immunol* 161: 3120–3127
15 Mehlhop PD, van de Rijn M, Brewer JP, Kisselgof AB, Geha RS, Oettgen HC, Martin TR (2000) CD40L, but not CD40, is required for allergen-induced bronchial hyperresponsiveness in mice. *Am J Respir Cell Mol Biol* 23: 646–651
16 Andre P, Prasad KS, Denis CV, He M, Papalia JM, Hynes RO, Phillips DR, Wagner DD (2002) CD40L stabilizes arterial thrombi by a beta3 integrin-dependent mechanism. *Nat Med* 8: 247–252
17 Prasad KS, Andre P, He M, Bao M, Manganello J, Phillips DR (2003) Soluble CD40 ligand induces beta3 integrin tyrosine phosphorylation and triggers platelet activation by outside-in signaling. *Proc Natl Acad Sci USA* 100: 12367–12371
18 Leveille C, Bouillon M, Guo W, Bolduc J, Sharif-Askari E, El-Fakhry Y, Reyes-Moreno C, Lapointe R, Merhi Y, Wilkins JA et al (2007) CD40 ligand binds to alpha5beta1 integrin and triggers cell signaling. *J Biol Chem* 282: 5143–5151
19 Zirlik A, Maier C, Gerdes N, MacFarlane L, Soosairajah J, Bavendiek U, Ahrens I, Ernst

S, Bassler N, Missiou A et al (2007) CD40 ligand mediates inflammation independently of CD40 by interaction with Mac-1. *Circulation* 115: 1571–1580

20 Noelle RJ, Ledbetter JA Aruffo A (1992) CD40 and its ligand, an essential ligand-receptor pair for thymus-dependent B-cell activation. *Immunol Today* 13: 431–433

21 Ramesh N, Fuleihan R, Geha R (1994) Molecular pathology of X-linked immunoglobulin deficiency with normal or elevated IgM (HIGMX-1). *Immunol Rev* 138: 87–104

22 Foy TM, Aruffo A, Bajorath J, Buhlmann JE, Noelle RJ (1996) Immune regulation by CD40 and its ligand GP39. *Annu Rev Immunol* 14: 591–617

23 Kawabe T, Naka T, Yoshida K, Tanaka T, Fujiwara H, Suematsu S, Yoshida N, Kishimoto T, Kikutani H (1994) The immune responses in CD40-deficient mice: impaired immunoglobulin class switching and germinal center formation. *Immunity* 1: 167–178

24 Renshaw BR, Fanslow WC 3rd, Armitage RJ, Campbell KA, Liggitt D, Wright B, Davison BL, Maliszewski CR (1994) Humoral immune responses in CD40 ligand-deficient mice. *J Exp Med* 180: 1889–1900

25 Rodriguez-Pinto D, Moreno J (2005) B cells can prime naive $CD4^+$ T cells *in vivo* in the absence of other professional antigen-presenting cells in a CD154-CD40-dependent manner. *Eur J Immunol* 35: 1097–1105

26 Caux C, Massacrier C, Vanbervliet B, Dubois B, Van Kooten C, Durand I, Banchereau J (1994) Activation of human dendritic cells through CD40 cross-linking. *J Exp Med* 180: 1263–1272

27 Cella M, Scheidegger D, Palmer-Lehmann K, Lane P, Lanzavecchia A, Alber G (1996) Ligation of CD40 on dendritic cells triggers production of high levels of interleukin-12 and enhances T cell stimulatory capacity: T-T help *via* APC activation. *J Exp Med* 184: 747–752

28 Hewitson JP, Hamblin PA, Mountford AP (2007) In the absence of CD154, administration of interleukin-12 restores Th1 responses but not protective immunity to *Schistosoma mansoni*. *Infect Immun* 75: 3539–3547

29 van Essen D, Kikutani H, Gray D (1995) CD40 ligand-transduced co-stimulation of T cells in the development of helper function. *Nature* 378: 620–623

30 Blotta MH, Marshall JD, DeKruyff RH, Umetsu DT (1996) Cross-linking of the CD40 ligand on human $CD4^+$ T lymphocytes generates a costimulatory signal that up-regulates IL-4 synthesis. *J Immunol* 156: 3133–3140

31 Amrani A, Serra P, Yamanouchi J, Han B, Thiessen S, Verdaguer J, Santamaria P (2002) CD154–dependent priming of diabetogenic $CD4(^+)$ T cells dissociated from activation of antigen-presenting cells. *Immunity* 16: 719–732

32 Bennett SR, Carbone FR, Karamalis F, Flavell RA, Miller JF, Heath WR (1998) Help for cytotoxic-T-cell responses is mediated by CD40 signalling. *Nature* 393: 478–480

33 Schoenberger SP, Toes RE, van der Voort EI, Offringa R, Melief CJ (1998) T-cell help for cytotoxic T lymphocytes is mediated by CD40–CD40L interactions. *Nature* 393: 480–483

34 Ridge JP, Di Rosa F, Matzinger P (1998) A conditioned dendritic cell can be a temporal bridge between a $CD4^+$ T-helper and a T-killer cell. *Nature* 393: 474–478

35 Bourgeois C, Rocha B, Tanchot C (2002) A role for CD40 expression on $CD8^+$ T cells in the generation of $CD8^+$ T cell memory. *Science* 297: 2060–2063

36 Schonbeck U, Libby P (2001) The CD40/CD154 receptor/ligand dyad. *Cell Mol Life Sci* 58: 4–43

37 Karmann K, Hughes CC, Schechner J, Fanslow WC, Pober JS (1995) CD40 on human endothelial cells: inducibility by cytokines and functional regulation of adhesion molecule expression. *Proc Natl Acad Sci USA* 92: 4342–4346

38 Karmann K, Min W, Fanslow WC, Pober JS (1996) Activation and homologous desensitization of human endothelial cells by CD40 ligand, tumor necrosis factor, and interleukin 1. *J Exp Med* 184: 173–182

39 Thienel U, Loike J, Yellin MJ (1999) CD154 (CD40L) induces human endothelial cell chemokine production and migration of leukocyte subsets. *Cell Immunol* 198: 87–95

40 Xu H, Arnaud F, Tadaki DK, Burkly LC, Harlan DM, Kirk AD (2001) Human platelets activate porcine endothelial cells through a CD154-dependent pathway. *Transplantation* 72: 1858–1861

41 Vowinkel T, Anthoni C, Wood KC, Stokes KY, Russell J, Gray L, Bharwani S, Senninger N, Alexander JS, Krieglstein CF et al (2007) CD40–CD40 ligand mediates the recruitment of leukocytes and platelets in the inflamed murine colon. *Gastroenterology* 132: 955–965

42 Zhou F, Ajuebor MN, Beck PL, Le T, Hogaboam CM, Swain MG (2005) CD154–CD40 interactions drive hepatocyte apoptosis in murine fulminant hepatitis. *Hepatology* 42: 372–380

43 Afford SC, Ahmed-Choudhury J, Randhawa S, Russell C, Youster J, Crosby HA, Eliopoulos A, Hubscher SG, Young LS, Adams DH (2001) CD40 activation-induced, Fas-dependent apoptosis and NF-kappaB/AP-1 signaling in human intrahepatic biliary epithelial cells. *FASEB J* 15: 2345–2354

44 Afford SC, Randhawa S, Eliopoulos AG, Hubscher SG, Young LS, Adams DH (1999) CD40 activation induces apoptosis in cultured human hepatocytes *via* induction of cell surface Fas ligand expression and amplifies fas-mediated hepatocyte death during allograft rejection. *J Exp Med* 189: 441–446

45 Shen XD, Ke B, Zhai Y, Amersi F, Gao F, Anselmo DM, Busuttil RW, Kupiec-Weglinski JW (2002) CD154-CD40 T-cell costimulation pathway is required in the mechanism of hepatic ischemia/reperfusion injury, and its blockade facilitates and depends on heme oxygenase-1 mediated cytoprotection. *Transplantation* 74: 315–319

46 Ke B, Shen XD, Gao F, Tsuchihashi S, Farmer DG, Briscoe D, Busuttil RW, Kupiec-Weglinski JW (2005) The CD154-CD40 T-cell co-stimulation pathway in liver ischemia and reperfusion inflammatory responses. *Transplantation* 79: 1078–1083

47 Moore TM, Shirah WB, Khimenko PL, Paisley P, Lausch RN, Taylor AE (2002) Involvement of CD40-CD40L signaling in postischemic lung injury. *Am J Physiol Lung Cell Mol Physiol* 283: L1255–1262

48 Ishikawa M, Vowinkel T, Stokes KY, Arumugam TV, Yilmaz G, Nanda A, Granger DN

(2005) CD40/CD40 ligand signaling in mouse cerebral microvasculature after focal ischemia/reperfusion. *Circulation* 111: 1690–1696

49 Parker DC, Greiner DL, Phillips NE, Appel MC, Steele AW, Durie FH, Noelle RJ, Mordes JP, Rossini AA (1995) Survival of mouse pancreatic islet allografts in recipients treated with allogeneic small lymphocytes and antibody to CD40 ligand. *Proc Natl Acad Sci USA* 92: 9560–9564

50 Larsen CP, Alexander DZ, Hollenbaugh D, Elwood ET, Ritchie SC, Aruffo A, Hendrix R, Pearson TC (1996) CD40-gp39 interactions play a critical role during allograft rejection. Suppression of allograft rejection by blockade of the CD40-gp39 pathway. *Transplantation* 61: 4–9

51 Larsen CP, Elwood ET, Alexander DZ, Ritchie SC, Hendrix R, Tucker-Burden C, Cho HR, Aruffo A, Hollenbaugh D, Linsley PS et al (1996) Long-term acceptance of skin and cardiac allografts after blocking CD40 and CD28 pathways. *Nature* 381: 434–438

52 Guo Z, Meng L, Kim O, Wang J, Hart J, He G, Alegre ML, Thistlethwaite JR Jr, Pearson TC, Larsen CP et al (2001) CD8 T cell-mediated rejection of intestinal allografts is resistant to inhibition of the CD40/CD154 costimulatory pathway. *Transplantation* 71: 1351–1354

53 Sun H, Subbotin V, Chen C, Aitouche A, Valdivia LA, Sayegh MH, Linsley PS, Fung JJ, Starzl TE, Rao AS (1997) Prevention of chronic rejection in mouse aortic allografts by combined treatment with CTLA4-Ig and anti-CD40 ligand monoclonal antibody. *Transplantation* 64: 1838–1843

54 Blazar BR, Taylor PA, Panoskaltsis-Mortari A, Buhlman J, Xu J, Flavell RA, Korngold R, Noelle R, Vallera DA (1997) Blockade of CD40 ligand-CD40 interaction impairs $CD4^+$ T cell-mediated alloreactivity by inhibiting mature donor T cell expansion and function after bone marrow transplantation. *J Immunol* 158: 29–39

55 Qian Y, Boisgerault F, Benichou G, Dana MR (2001) Blockade of CD40-CD154 costimulatory pathway promotes survival of allogeneic corneal transplants. *Invest Ophthalmol Vis Sci* 42: 987–994

56 Tung TH, Mackinnon SE, Mohanakumar T (2003) Long-term limb allograft survival using anti-CD40L antibody in a murine model. *Transplantation* 75: 644–650

57 Hancock WW, Sayegh MH, Zheng XG, Peach R, Linsley PS, Turka LA (1996) Costimulatory function and expression of CD40 ligand, CD80, and CD86 in vascularized murine cardiac allograft rejection. *Proc Natl Acad Sci USA* 93: 13967–13972

58 Hancock WW, Buelow R, Sayegh MH, Turka LA (1998) Antibody-induced transplant arteriosclerosis is prevented by graft expression of anti-oxidant and anti-apoptotic genes. *Nat Med* 4: 1392–1396

59 Iwakoshi NN, Mordes JP, Markees TG, Phillips NE, Rossini AA, Greiner DL (2000) Treatment of allograft recipients with donor-specific transfusion and anti-CD154 antibody leads to deletion of alloreactive $CD8^+$ T cells and prolonged graft survival in a CTLA4-dependent manner. *J Immunol* 164: 512–521

60 Honey K, Cobbold SP, Waldmann H (1999) CD40 ligand blockade induces $CD4^+$ T cell tolerance and linked suppression. *J Immunol* 163: 4805–4810

61 Jones ND, Van Maurik A, Hara M, Spriewald BM, Witzke O, Morris PJ, Wood KJ (2000) CD40-CD40 ligand-independent activation of CD8+ T cells can trigger allograft rejection. *J Immunol* 165: 1111–1118

62 Trambley J, Bingaman AW, Lin A, Elwood ET, Waitze SY, Ha J, Durham MM, Corbascio M, Cowan SR, Pearson TC et al (1999) Asialo GM1(+) CD8(+) T cells play a critical role in costimulation blockade-resistant allograft rejection. *J Clin Invest* 104: 1715–1722

63 Corbascio M, Mahanty H, Osterholm C, Qi Z, Pearson TC, Larsen CP, Freise CE, Ekberg H (2002) Anti-lymphocyte function-associated antigen-1 monoclonal antibody inhibits CD40 ligand-independent immune responses and prevents chronic vasculopathy in CD40 ligand-deficient mice. *Transplantation* 74: 35–41

64 Nicolls MR, Coulombe M, Beilke J, Gelhaus HC, Gill RG (2002) CD4–dependent generation of dominant transplantation tolerance induced by simultaneous perturbation of CD154 and LFA-1 pathways. *J Immunol* 169: 4831–4839

65 Sho M, Harada H, Rothstein DM, Sayegh MH (2003) CD45RB-targeting strategies for promoting long-term allograft survival and preventing chronic allograft vasculopathy. *Transplantation* 75: 1142–1146

66 Molano RD, Pileggi A, Berney T, Poggioli R, Zahr E, Oliver R, Ricordi C, Rothstein DM, Basadonna GP, Inverardi L (2003) Prolonged islet allograft survival in diabetic NOD mice by targeting CD45RB and CD154. *Diabetes* 52: 957–964

67 Ozkaynak E, Gao W, Shemmeri N, Wang C, Gutierrez-Ramos JC, Amaral J, Qin S, Rottman JB, Coyle AJ, Hancock WW (2001) Importance of ICOS-B7RP-1 costimulation in acute and chronic allograft rejection. *Nat Immunol* 2: 591–596

68 Gao W, Demirci G, Strom TB, Li XC (2003) Stimulating PD-1–negative signals concurrent with blocking CD154 co-stimulation induces long-term islet allograft survival. *Transplantation* 76: 994–999

69 Smiley ST, Csizmadia V, Gao W, Turka LA, Hancock WW (2000) Differential effects of cyclosporine A, methylprednisolone, mycophenolate, and rapamycin on CD154 induction and requirement for NFkappaB: implications for tolerance induction. *Transplantation* 70: 415–419

70 Sho M, Sandner SE, Najafian N, Salama AD, Dong V, Yamada A, Kishimoto K, Harada H, Schmitt I, Sayegh MH (2002) New insights into the interactions between T-cell costimulatory blockade and conventional immunosuppressive drugs. *Ann Surg* 236: 667–675

71 Blaha P, Bigenzahn S, Koporc Z, Schmid M, Langer F, Selzer E, Bergmeister H, Wrba F, Kurtz J, Kiss C et al (2003) The influence of immunosuppressive drugs on tolerance induction through bone marrow transplantation with costimulation blockade. *Blood* 101: 2886–2893

72 Lombardi G, Sidhu S, Daly M, Batchelor JR, Makgoba W, Lechler RI (1990) Are primary alloresponses truly primary? *Int Immunol* 2: 9–13

73 Valujskikh A, Pantenburg B, Heeger PS (2002) Primed allospecific T cells prevent the

effects of costimulatory blockade on prolonged cardiac allograft survival in mice. *Am J Transplant* 2: 501
74 Zhai Y, Meng L, Gao F, Busuttil RW, Kupiec-Weglinski JW (2002) Allograft rejection by primed/memory CD8+ T cells is CD154 blockade resistant: therapeutic implications for sensitized transplant recipients. *J Immunol* 169: 4667–4673
75 Wu Z, Bensinger SJ, Zhang J, Chen C, Yuan X, Huang X, Markmann JF, Kassaee A, Rosengard BR, Hancock WW et al (2004) Homeostatic proliferation is a barrier to transplantation tolerance. *Nat Med* 10: 87–92
76 Vu MD, Clarkson MR, Yagita H, Turka LA, Sayegh MH, Li XC (2006) Critical, but conditional, role of OX40 in memory T cell-mediated rejection. *J Immunol* 176: 1394–1401
77 Braciale TJ, Andrew ME, Braciale VL (1981) Simultaneous expression of H-2-restricted and alloreactive recognition by a cloned line of influenza virus-specific cytotoxic T lymphocytes. *J Exp Med* 153: 1371–1376
78 Yang H, Welsh RM (1986) Induction of alloreactive cytotoxic T cells by acute virus infection of mice. *J Immunol* 136: 1186–1193
79 Adams AB, Williams MA, Jones TR, Shirasugi N, Durham MM, Kaech SM, Wherry EJ, Onami T, Lanier JG, Kokko KE et al (2003) Heterologous immunity provides a potent barrier to transplantation tolerance. *J Clin Invest* 111: 1887–1895
80 Pantenburg B, Heinzel F, Das L, Heeger PS, Valujskikh A (2002) T cells primed by *Leishmania major* infection cross-react with alloantigens and alter the course of allograft rejection. *J Immunol* 169: 3686–3693
81 Welsh RM, Markees TG, Woda BA, Daniels KA, Brehm MA, Mordes JP, Greiner DL, Rossini AA (2000) Virus-induced abrogation of transplantation tolerance induced by donor-specific transfusion and anti-CD154 antibody. *J Virol* 74: 2210–2218
82 Forman D, Welsh RM, Markees TG, Woda BA, Mordes JP, Rossini AA, Greiner DL (2002) Viral abrogation of stem cell transplantation tolerance causes graft rejection and host death by different mechanisms. *J Immunol* 168: 6047–6056
83 Thornley TB, Brehm MA, Markees TG, Shultz LD, Mordes JP, Welsh RM, Rossini AA, Greiner DL (2006) TLR agonists abrogate costimulation blockade-induced prolongation of skin allografts. *J Immunol* 176: 1561–1570
84 Chen L, Wang T, Zhou P, Ma L, Yin D, Shen J, Molinero L, Nozaki T, Phillips T, Uematsu S et al (2006) TLR engagement prevents transplantation tolerance. *Am J Transplant* 6: 2282–2291
85 Zhai Y, Shen XD, Gao F, Coito AJ, Wasowska BA, Salama A, Schmitt I, Busuttil RW, Sayegh MH, Kupiec-Weglinski JW (2002) The CD154-CD40 T cell costimulation pathway is required for host sensitization of CD8(+) T cells by skin grafts *via* direct antigen presentation. *J Immunol* 169: 1270–1276
86 Zhai Y, Meng L, Busuttil RW, Sayegh MH, Kupiec-Weglinski JW (2003) Activation of alloreactive CD8+ T cells operates *via* CD4-dependent and CD4-independent mechanisms and is CD154 blockade sensitive. *J Immunol* 170: 3024–3028
87 Gao D, Lunsford KE, Eiring AM, Bumgardner GL (2004) Critical role for CD8 T cells

in allograft acceptance induced by DST and CD40/CD154 costimulatory blockade. *Am J Transplant* 4: 1061–1070

88 Wells AD, Li XC, Li Y, Walsh MC, Zheng XX, Wu Z, Nunez G, Tang A, Sayegh M, Hancock WW et al (1999) Requirement for T-cell apoptosis in the induction of peripheral transplantation tolerance. *Nat Med* 5: 1303–1307

89 Li Y, Li XC, Zheng XX, Wells AD, Turka LA, Strom TB (1999) Blocking both signal 1 and signal 2 of T-cell activation prevents apoptosis of alloreactive T cells and induction of peripheral allograft tolerance. *Nat Med* 5: 1298–1302

90 Sanchez-Fueyo A, Domenig C, Strom TB, Zheng XX (2002) The complement dependent cytotoxicity (CDC) immune effector mechanism contributes to anti-CD154 induced immunosuppression. *Transplantation* 74: 898–900

91 Monk NJ, Hargreaves RE, Marsh JE, Farrar CA, Sacks SH, Millrain M, Simpson E, Dyson J, Jurcevic S (2003) Fc-dependent depletion of activated T cells occurs through CD40L-specific antibody rather than costimulation blockade. *Nat Med* 9: 1275–1280

92 Nathan MJ, Yin D, Eichwald EJ, Bishop DK (2002) The immunobiology of inductive anti-CD40L therapy in transplantation: allograft acceptance is not dependent upon the deletion of graft-reactive T cells. *Am J Transplant* 2: 323–332

93 Taylor PA, Noelle RJ, Blazar BR (2001) CD4($^+$)CD25($^+$) immune regulatory cells are required for induction of tolerance to alloantigen *via* costimulatory blockade. *J Exp Med* 193: 1311–1318

94 Taylor PA, Friedman TM, Korngold R, Noelle RJ, Blazar BR (2002) Tolerance induction of alloreactive T cells *via ex vivo* blockade of the CD40:CD40L costimulatory pathway results in the generation of a potent immune regulatory cell. *Blood* 99: 4601–4609

95 Zhai Y, Meng L, Gao F, Busuttil RW, Kupiec-Weglinski JW (2002) Allograft rejection by primed/memory CD8(+) T cells is CD154 blockade resistant: therapeutic implications for sensitized transplant recipients. *J Immunol* 169: 4667–4673

96 Quezada SA, Bennett K, Blazar BR, Rudensky AY, Sakaguchi S, Noelle RJ (2005) Analysis of the underlying cellular mechanisms of anti-CD154-induced graft tolerance: the interplay of clonal anergy and immune regulation. *J Immunol* 175: 771–779

97 Kurtz J, Shaffer J, Lie A, Anosova N, Benichou G, Sykes M (2004) Mechanisms of early peripheral CD4 T-cell tolerance induction by anti-CD154 monoclonal antibody and allogeneic bone marrow transplantation: evidence for anergy and deletion but not regulatory cells. *Blood* 103: 4336–4343

98 Linhart B, Bigenzahn S, Hartl A, Lupinek C, Thalhamer J, Valenta R, Wekerle T (2007) Costimulation blockade inhibits allergic sensitization but does not affect established allergy in a murine model of grass pollen allergy. *J Immunol* 178: 3924–3931

CTLA-4-immunoglobulin and indoleamine 2,3-dioxygenase in dominant tolerance

Francesca Fallarino, Carmine Vacca, Claudia Volpi, Maria T. Pallotta, Stefania Gizzi, Ursula Grohmann and Paolo Puccetti

Department of Experimental Medicine, University of Perugia, 06126 Perugia, Italy

Regulatory T cells and tolerogenic antigen-presenting cells in dominant tolerance

The immune system is delicately balanced between self-antigen-driven tolerance and pathogen-driven immunity. In the healthy individual, these two states represent a sliding scale of responsiveness. A shift toward the extreme ends of this scale, i.e., lack of response or an excessive response (such as in autoimmunity and allergy) results in pathophysiological conditions that may be at the basis of diseases. As a consequence, several immune mechanisms have evolved to protect against T and B cells harboring the potential to recognize and become activated by self antigens. Establishment and regulation of self tolerance are exerted at two levels. First, the so-called "central tolerance", which allows selection of T cells in the thymus (where the gene *AIRE* permits expression of tissue-specific genes), takes place during T cell development, and contributes to preventing maturation of autoreactive T lymphocytes [1–4]. In this process, the majority of self-reactive T cells are deleted by a mechanism termed "negative selection", but at the same time, some $CD4^+$ T cells differentiate to the $CD4^+CD25^+$Foxp3-expressing regulatory T cell (Treg) lineage [5–7]. The parameters specifying whether autoreactive $CD4^+$ thymocytes are deleted (recessive tolerance) or differentiate into Tregs (dominant tolerance) remain unclarified.

Second, the fact that autoreactive T cells have been identified in both animals and humans indicates that negative selection alone is not sufficient to prevent autoimmunity [8]. Therefore, additional mechanisms contribute to regulating the activation of mature T cells in lymphoid or nonlymphoid organs (peripheral tolerance) [9, 10]. These mechanisms include (i) ignorance due to exclusion of T cells from the site of antigen expression or low avidity of their T cell receptors (TCRs) for self antigen; (ii) tolerance mediated by $CD4^+CD25^+$ Tregs; and (iii) deletion or inactivation of self-reactive T cells. It is now recognized that a network of both Tregs and tolerogenic antigen-presenting cells (APCs) exists that, engaged in the dominant control

The Immune Synapse as a Novel Target for Therapy, edited by Luis Graca

of self-reactive T cell responses, down-regulate immunity in various inflammatory circumstances and assure peripheral T cell tolerance.

It is now firmly established that there are both "natural" (or constitutive) and "inducible" (or adaptive) populations of Tregs, which probably have complementary and overlapping functions in the control of immune responses [11]. Natural Tregs express the cell-surface marker CD25; however, CD25 is not a specific marker of natural Tregs. CD25 is an activation marker for T cells and is, therefore, also expressed by effector T helper (Th) 1 and Th2 cells, and suppressive function has also been documented for $CD25^-$ T cells. Other putative markers for Tregs include cell-surface expression of CD38, CD62L, CD103 or glucocorticoid-induced tumor-necrosis factor (TNF) receptor (GITR), or low levels of cell-surface CD45RB expression, and importantly cytotoxic T lymphocyte-associated antigen-4 (CTLA-4) or intracellular expression of the transcriptional repressor FOXP3 (forkhead box P3) [12–14]. At the moment, this factor seems to be the most promising marker of natural Tregs, and studies have shown that transfection of $CD4^+CD25^-$ T cells with *Foxp3* provides them with intracellular regulatory activity [15]. Other populations of antigen-specific Tregs can be induced from naive $CD4^+CD25^-$ or $CD8^+CD25^-$ T cells in the periphery under the influence of specific dendritic cells (DCs), interleukin-10 (IL-10), transforming growth factor-β (TGF-β) and possibly interferon-γ (IFN-γ), and they include different populations of $CD4^+$ T cell such as T regulatory 1 (TR1) cells, and Th3 cells [16]. Although $CD8^+$ T cells are normally associated with cytotoxic T lymphocyte function and IFN-γ production, these cells, or subtypes thereof, can secrete IL-10 and have been called "$CD8^+$ regulatory T cells" [17, 18]. In addition, natural killer T (NKT) cells, which co-express NK-cell and T-cell markers, can secrete regulatory cytokines, including IL-10 [19].

The mechanism of the suppressive function of natural and inducible Tregs is still debated, but in different model systems, suppressive activity has been shown to be multifactorial and includes either the secretion of cytokines or cell-cell contact [20, 21]. In particular, the latter seems to be critical for the inhibitory activity of $CD4^+CD25^+$ Tregs, and it has been shown that the expression of the inhibitory costimulatory molecule CTLA-4 might be involved [22]. However, there is also conflicting evidence concerning the roles of IL-10 and secreted or cell-surface TGF-β [23, 24].

In addition to Tregs, recent studies point to an important role for DCs in the induction of peripheral tolerance and in the regulation of central tolerance [25, 26]. DCs are specialized APCs critical to initiating T cell responses *in vitro* and *in vivo* and which play a pivotal role in the induction of both immunity and tolerance [27]. The selection between these two opposing outcomes seems to depend on DC maturity and/or activation [28]. Moreover, DC heterogeneity and their high degree of plasticity also contribute to the dichotomous function of DCs [29]. Moreover, all these distinct variables can be affected by the *in vivo* microenvironment that interferes with the properties and migration of DCs, resulting in the induction of specific immune responses. These features of DCs relate to their ability to integrate a diverse array of signals and then direct an appropriate immune response [30]. Several factors can drive DC maturation and activation toward the induction of immunity or tolerance, including cytokine milieu (i.e., IL-10 and TGF-β), pathogen-associated molecular pattern-containing components of bacteria and viruses, such as lipopolysaccharide and CpG-rich motifs, and activation of specific costimulatory molecules, such as those in the B7 family, which are bound by specific molecules expressed on T cells.

Interestingly, among these factors, some have been reported to induce the expression and activity of the tolerogenic enzyme indoleamine 2,3-dioxygenase (IDO) in specific DC subsets. There is an increasing appreciation of the unifying role of IDO in mediating tolerance under a variety of physiopathological conditions, together with a growing recognition of the effect of IDO activation in DC subsets (i.e., plasmacytoid DCs, pDCs) as one important cell-cell contact-dependent mechanism of action of Tregs that express surface CTLA-4 [31, 32].

IDO in dominant tolerance

IDO modulation in APCs during immune responses

IDO, a tryptophan-degrading enzyme, has recently emerged as an important immunomodulator of T cell function and an inducer of tolerance [32, 33]. IDO, encoded by the human *INDO* and murine *Indo* genes located in the short arm of chromosome 8, regulates immune responses by suppressing effector T cell function through its capacity to catabolize the essential amino acid tryptophan and by production of neuroactive and immunoregulatory metabolites known as kynurenines (Fig. 1). The *INDO* gene has been conserved over the last 600 million years of evolution, and during adaptation, this gene could have served as a bridge between innate and adaptive immunity, participating in the antibacterial defense by macrophages and in the modulation of T cells by DCs [34]. The gene encodes an inducible heme-monooxygenase and the crystal structure of the human IDO has been recently characterized [35]. The biological relevance of IDO-mediated tryptophan catabolism to peripheral immune tolerance was first shown by the work by Munn and Mellor [36], which revealed an important role for cells expressing IDO in regulating maternal T cell immunity during pregnancy. Blocking IDO by a small molecule inhibitor of IDO, 1-methyl-tryptophan (1-MT), during murine pregnancy resulted in allogeneic fetal rejection. These findings were consistent with *in vitro* results indicating that degradation of tryptophan by cultured macrophages blocks T cell activation.

DCs have the ability to stimulate naive T cells but are also involved in tolerance [37]. Interestingly, IDO can be highly modulated in DC subsets [38]. IDO activity is regulated at both transcriptional and post-transcriptional levels and IFN-γ

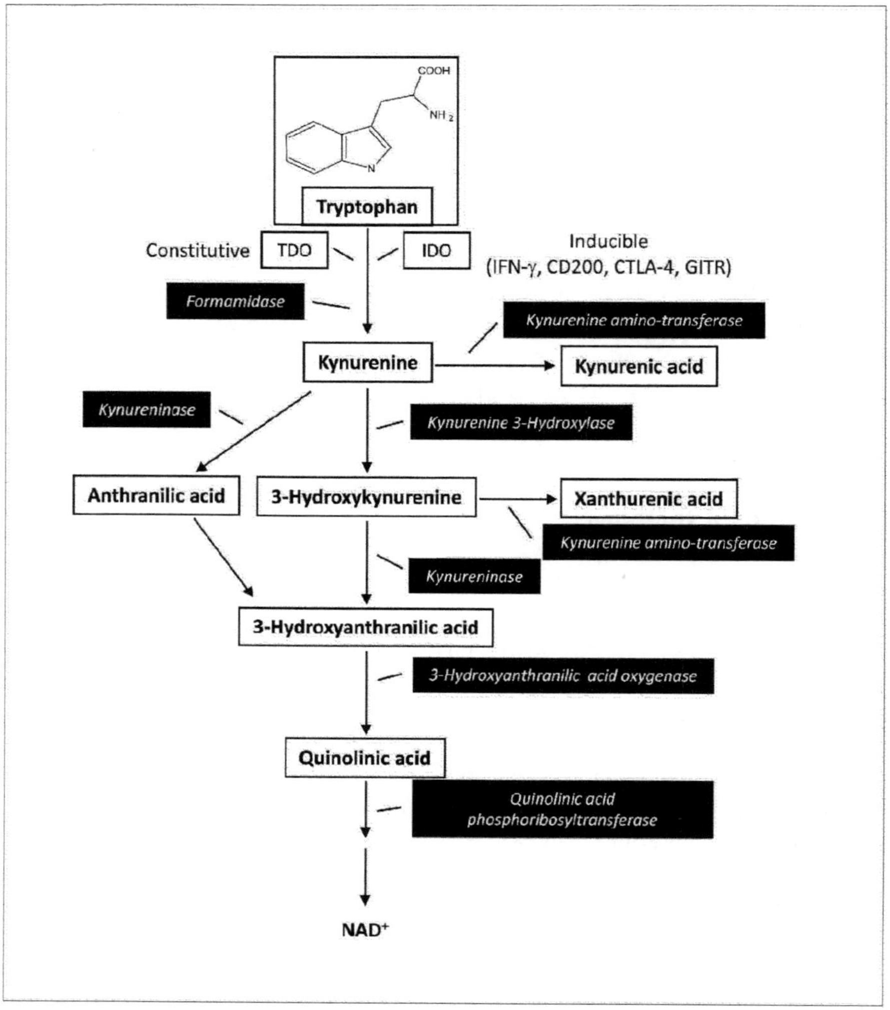

Figure 1
Tryptophan catabolic pathways.
Tryptophan is metabolized along the kynurenine pathway, in which inducible indoleamine 2,3-dioxygenase (IDO) and constitutive tryptophan 2,3-dioxygenase (TDO) catalyze the rate-limiting step of the respective pathways. Each enzyme catalyzes the same reaction, the oxidative cleavage of the 2,3 double bond in the indole ring. IDO expression is inducible in cells of the immune system and is subjected to complex regulation by a variety of immunological signals.

represents the principal regulator of *Indo* transcription [39, 40]. We and others have shown that IFN-γ is a potent inducer of IDO activity both in murine $CD8^+$ and human DC subsets [41, 42]. Interestingly, the same cytokine, IFN-γ, was not able to confer tryptophan-catabolizing activity on $CD8^-$ DCs, although the resultant IDO protein expression was comparable in the two DC subsets. Beside IFN-γ, several additional stimuli have been reported to influence both IDO expression and functional activity in specific target cells and they include specific Toll-like receptor (TLR) ligands and costimulatory molecules. The specific TLR9 ligands CpG oligodeoxynucleotides and thymosin-α1 and the TLR4 ligand LPS have been reported to affect IDO expression. Interestingly, CpG and thymosin-α1 were able alone to modulate IDO *in vivo* and *in vitro*, while LPS stimulated IDO expression only in combination with TNF-α [43–45].

In addition to cytokines and TLRs, engagement of costimulatory molecules on conventional DCs (cDCs) or pDCs also results in IDO activation. In particular, our group has demonstrated that pDCs ($CD11c^+$ mPDCA-1^+ $120G8^+$) do not express functional IDO constitutively, but tryptophan catabolism can be activated by tolerogenic ligands. Freshly isolated pDCs express low levels of costimulatory molecules, including B7-1 and B7-2, yet both CTLA-4-Ig and CD28-Ig are capable of binding those molecules and inducing IDO activation [46]. Interestingly, tryptophan conversion to kynurenine by CTLA-4-Ig required B7-1, similar to that observed with cDCs [47]. These data suggest that engagement of B7 molecules on DCs by CTLA-4-expressing cells may represent an important mechanism of suppression by Tregs [31]. CD28-Ig, which is able to potentiate the immunogenic properties of DCs, is also capable of activating IDO functional activity in DCs when the function of the suppressor of cytokine signaling 3 (SOCS3) is impaired. In fact, we have previously shown that CD28-Ig up-regulates SOCS3 and this factor is required to prevent the IFN-γ-driven induction of immunosuppressive IDO; conversely, in the absence of SOCS3, in both cDCs and pDCs, CD28-Ig promotes release of IFN-γ, activation of IDO and the emergence of a tolerogenic phenotype that is sustained by both IFN-γ and IL-6 [48]. In addition to the engagement of B7 molecules, activation of CD200R1 by CD200-Ig in conventional DCs and pDCs activates tolerogenic programs, with a predominance of IDO-dependent effects in pDCs [46]. 4-1BB, a member of the TNF receptor superfamily, has been reported to activate IDO; the administration of an anti-4-1BB agonist antibody allowed IDO expression to be induced in APCs in a model of rheumatoid arthritis, and this prevented the development of the disease in this model [49]. Recently, we have demonstrated that engagement of another member of the TNF superfamily, GITRL, in pDCs by a soluble form of GITR protein (GITR-Ig) also activates tolerogenic programs, characterized by IDO-dependent effects. As opposed to CTLA-4-Ig, where type II IFNs are required for IDO induction, this study indicates that the autocrine production of type I IFNs is crucially involved in IDO activation by reverse signaling through GITR-Ig. Interestingly, dexamethasone administered *in vivo* activated IDO through the concomitant induction of GITR in CD4+ T cells and GITRL in pDCs, and this mechanism contributed to protection against allergic bronchopulmonary aspergillosis [50].

Mechanisms of IDO-mediated suppression of immunity

The downstream molecular mechanisms by which IDO affects immune responses are still a matter of active investigation. The possibilities include direct effects on T cells or APCs, mediated either by tryptophan depletion or by downstream tryptophan catabolites, named kynurenines, the main products of tryptophan catabolism by the inducible enzyme IDO (Fig. 1). Several studies have provided definite evidence for an important role of kynurenine metabolites in IDO-mediated modulation of immune functions. Based on recent data, it is possible to assume that tryptophan-derived kynurenines affect immune responses by at least three different mechanisms: induction of T cell apoptosis; promotion of regulatory cell activity; and modulation of specific effector functions in selected populations of immune cells such as CD4+ and CD8+ T cells, DCs and NK cells [51–54]. Concerning the pro-apoptotic effects, our group has demonstrated that the kynurenine derivatives 3-hydroxyanthranilic acid (3-HAA) and quinolinic acid can induce selective apoptosis *in vitro* and *in vivo* of murine thymocytes and of antigen-specific CD4+ Th1 more than Th2 cells [55].

Interestingly, over the years, two major theories have been proposed to explain how tryptophan catabolism creates tolerance. The first theory assumes that tryptophan breakdown suppresses T cell proliferation by dramatically reducing the supply of this essential amino acid in local tissue microenvironments [32]. The second theory postulates that the downstream metabolites of tryptophan catabolism act to suppress immune cells by mediating pro-apoptotic effects [31]. In addition to these mechanisms, we have recently demonstrated that both tryptophan starvation and tryptophan catabolites contribute to establishing a regulatory environment affecting CD8+ as well as CD4+ T cell function, leading to *de novo* generation of Foxp3-expressing Tregs [56]. The effect was demonstrable *in vivo* and required, similar to Foxp3 induction, the activity of GCN2 kinase; these cells showed a reduced cytotoxic activity *in vitro*. Interestingly, in an independent study, it had previously been reported that the Treg phenotype could result, in part, from a biochemical reaction involving the modulatory activity of the tryptophan catabolite, 3-HAA [57]. These data suggest that IDO-induced mechanisms are not mutually exclusive, each possibly operating in a specific cell type and all concurring to immunomodulation by the enzyme [58]. Moreover, the relative contributions of tryptophan depletion and the production of toxic metabolites to IDO-dependent inhibition of T cell responses could vary according to the type of APCs that are mediating the effect. Reconciling these different possibilities, however, could be crucial to understanding how tryptophan catabolism is involved in mediating aspects of immune regulation that are both differentiated and interdependent.

The protein CTLA-4 in dominant tolerance

Functional expression of CTLA-4 in immune cells

T cells play a central role in the initiation and regulation in the adaptive immune response to antigens, and naive T cells require more than one signal for their full activation [59]. The first, signal 1, is an antigen-specific signal provided by the TCR interacting with the major histocompatibility complex (MHC) and antigenic peptide complex on the APC. The second, or costimulatory signal, is provided by the interactions between specific receptors on the T cell and their ligands on the APC. Following these two signals, a number of pathways are activated on the T cells and by reverse signaling on the APCs, with subsequent transcription of a number of effector factors, cytokines and other costimulatory/inhibitory molecules. These factors are responsible, firstly, for T cell proliferation, generation of an effector CD4$^+$ T cell pool (T helper), clonal expansion of activated CD8$^+$ cytotoxic T cells and further APC maturation, and, secondly, for the regulation of the same T cell effector function [60]. In the absence of the second signal, a state of T cell unresponsiveness (anergy) can result, which is also responsible for decreased T cell responses after further T cell stimulation.

Multiple costimulatory pathways are involved in T cell regulation these can either up-regulate or down-regulate T cell activation [61]. Perhaps the most critical and best-characterized costimulatory/co-inhibitory interaction is between CD28 or CTLA-4 and CD80 (B7-1) and CD86 (B7-2) molecules in the B7 family [61, 62]. CD28/B7 family members are type I transmembrane glycoproteins and members of the Ig superfamily. CD28 is constitutively expressed in T cells and its interaction with its ligands CD80 and CD86 is essential for initiating antigen-specific T cell responses [63]. Both B7-1 and B7-2 are expressed on a variety of APCs, including DCs, Langerhans cells, activated macrophages and B cells, but they have distinct kinetics of expression. B7-2 is constitutively expressed at low levels on resting APCs; B7-1 is instead generally absent [64]. Both molecules undergo marked up-regulation upon cellular activation, and further induction of these molecules has been reported upon CD40 ligation and cytokine receptor signaling by IFN-γ and IL-4 [63]. Both CD80 and CD86 bind to a second receptor on T cell, CTLA-4, which shares ~30% homology with CD28 [65]. CTLA-4 is a higher affinity receptor for B7-1 and B7-2 than CD28, binding both molecules ~500-fold more avidly [66, 67]. It is induced at 24–48 h after T cell activation in both naive and primed CD4$^+$ and CD8$^+$ T cells, but it is constitutively expressed on natural Tregs and in memory T cells [68–70]. The transcriptional regulation of *CTLA4* gene expression is only partially known and may be dependent on NFAT (nuclear factor of activated T cells) because modulation of NFAT levels correlates directly with CTLA-4 expression [71]. The evolutionary conservation of CTLA-4, and the differential expression in T cell subsets suggest that transcriptional and translational changes in its expression may have implications in immunological responses. To date, four main polymorphisms of the *CTLA4* gene have been identified and studied in the context of autoimmune disorders. The A49G polymorphism is the only one that changes the primary amino acid sequence of CTLA-4 and *in vitro* studies suggest that this mutant form is differently processed in the endoplasmic reticulum, leading to reduced surface expression [72].

By crystallography and modeling, tertiary structures for ligands and receptors superfamily have been determined. CD28 is expressed as a homodimer, and interacts with B7 ligands through an MYPPPY recognition motif in its extracellular domain [73]. CTLA-4 also interacts with B7 molecules by virtue of the MYPPPY sequence, but structural differences allow CTLA-4 homodimers to bind B7 from two different dimers forming a lattice-like structure and this may explain CTLA-4 higher avidity for B7 molecules [74, 75]. Interaction of CTLA-4 with the APCs allows accumulation of CTLA-4 at the immune synapse and this requires B7-1 and little contribution of B7-2 [76, 77]. In addition, although B7-1 and B7-2 can both contribute to positive and negative costimulation (through CD28 and CTLA-4, respectively), B7-2 could be more important for the initiation of an immune response through the interaction with CD28, while B7-1 may preferentially mediate CTLA-4-derived inhibitory effects [48, 78]. Interestingly, the effects of CD28 costimulation and CTLA-4 inhibition are uncoupled in E3-ubiquitin ligase Cbl-b-deficient T cells, suggesting that the function of CD28 and CTLA-4 depends on the expression of this enzyme [79, 80]. This distinction in binding preference is important when considering therapeutic intervention for the prevention of allograft rejection (B7-2-mediated initial activation) or for reducing inflammation in autoimmune diseases, such as rheumatoid arthritis (CD80-mediated inhibition). Although new members of the B7 family have been identified, only B7-1 and B7-2 behave as true endogenous ligands for CTLA-4, although reports point to alternative forms of CTLA-4 characterized by B7-independent effects [81, 82].

The positive role of CD28 during T cell activation involves enhanced proliferation and increased T cell survival by augmenting the production of several cytokines such as IL-2 and its receptor and anti-apoptotic molecules, such as BcL-x_L [83–85]. In addition to a direct effect of CD28 signaling on T cell activation and differentiation, more recent data suggest that CD28 may trigger reverse B7-mediated signals leading to enhanced DC stimulatory functions [47].

In contrast to CD28, the inhibitory nature of CTLA-4 was confirmed with the generation of CTLA-4-deficient mice that develop severe lymphoproliferation and autoimmunity and die a few weeks after birth, showing that lack of CTLA-4 results in loss of tolerance [86, 87]. The lack of CTLA-4 impairs the ability of T cells to undergo cell-cycle arrest after activation, resulting in progressive accumulation of activated T cells that gradually infiltrate vital organs and cause death [88]. Such a robust proliferation is not due to a defect in T cell apoptosis and this phenotype suggests a generalized expansion of self-reactive T cells that are no longer held in check, suggesting that CTLA-4 is involved not only in negative regulation of T cell activation but also in the maintenance of T cell homeostasis [89]. The thymus of CTLA-4-deficient mice remains normal in size; alterations in thymocyte populations are not detected until later stages of life and negative selection seems mostly normal, although CTLA-4 may fine-tune negative selection [90]. An observation that is still puzzling is that the CTLA-4 deficiency in T cells is not cell autonomous. Interestingly, the phenotype of CTLA-4-deficient mice was shown to be dependent on B7 engagement, in the absence of CTLA-4, by the other ligand CD28 [91], and either transgenic expression of CTLA-4 (full length or a tail-less/truncated form) or administration of the soluble form of CTLA-4 (CTLA-4-Ig) *in vivo* was capable of rescuing the phenotype of CTLA-4-deficient mice [90, 92].

Three major mechanisms may explain the inhibitory function of CTLA-4 *in vivo* as recently reviewed [43]. These mechanisms involve competitive antagonisms of CTLA-4 of CD28-induced signals; direct CTLA-4 negative signal on T cells resulting in the inhibition of both early as well as probably downstream TCR-mediated signals on T effector and/or Tregs; and *via* reverse signaling in B7-expressing APCs [31, 93, 94]. Regarding the latter, exposure of murine B7-expressing DCs to CTLA-4 transfected or endogenously expressing Tregs resulted in activation of IDO enzyme activity in DCs; this effect involved activation of the NF-κB pathway and required the production of type II IFNs (IFN-γ), but did not require CTLA-4 signaling on T cells [31]. In fact, higher IDO induction was detected by co-incubation of DCs with Jurkat-transfected T cells with a construct encoding for tail-less CTLA-4, which cannot signal but is stably retained at the cell surface [31]. Several sets of data suggest that these mechanisms *in vivo* may not be mutually exclusive and probably they all work in synergy to maintain immune homeostasis.

CTLA-4-Ig as a means to mimic specific effects of CTLA-4 and the central role of IDO induction

Because the inhibitory effect of CTLA-4 on immune responses is so potent, several approaches have been taken to target CTLA-4 for clinical applications. An interesting therapeutic application using CTLA-4 has been the creation of the CTLA-4-Ig fusion protein. This protein is comprised of an extracellular domain of human or mouse CTLA-4 fused to the hinge, CH2 and CH3 domains of a human or mouse IgG (Fig. 2). In addition to blocking CD28 engagement, CTLA-4-Ig, like native CTLA-4 on the membrane of Tregs, has been shown to bind to B7 and induce reverse signaling to DCs, resulting in IDO activation in these cells [95]. The immunosuppressive effect of CTLA-4-Ig in a diabetic mouse model of pancreatic β-cell transplantation was shown to depend on tryptophan catabolism as it could be prevented by administration of the specific competitive inhibitor of IDO (1-MT). The activation of IDO by CTLA-4-Ig may contribute to establishing a tolerogenic

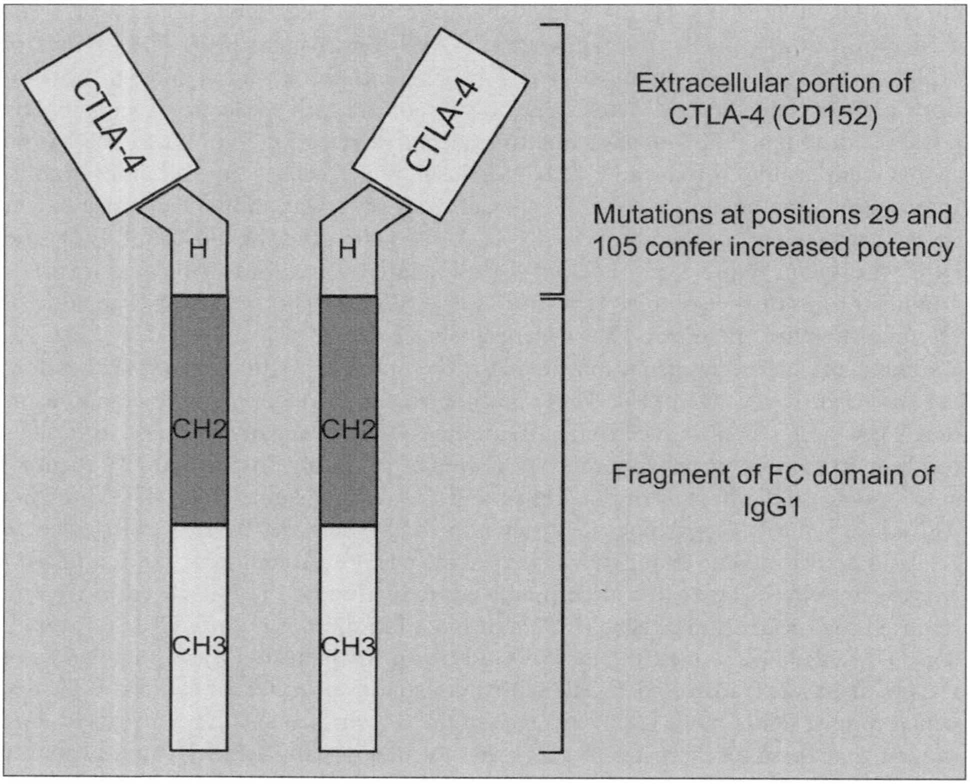

Figure 2
Structure of CTLA-4-Ig.
Abatacept is a fusion protein comprised of the extracellular domain of CTLA4 with the hinge (H), CH2 and CH3 domains of IgG1. Belatacept is the basis for second-generation biodrugs of this type and is currently being tested in clinical trials. It contains two amino acid substitutions (L104E and A29Y) responsible for slower dissociation rates for both CD86 and CD80.

network of cells that contribute to reinforce IDO induction and probably Treg generation as depicted in Figure 3.

In fact, deregulation of tryptophan catabolism has been associated with various types of diseases with immunological components, such as tumors [96, 97] as well autoimmune disorders [98, 99], allergic responses and infections [100]. Several cell types have been implicated in establishing this tolerant state, including Tregs [6], Tr1 cells [101] and myeloid cells [102]. By expressing specific inhibitory molecules, these cell types become able to activate the IDO tolerogenic pathway; one of the best-studied inhibitory molecules able to mediate this effect is CTLA-4 [31].

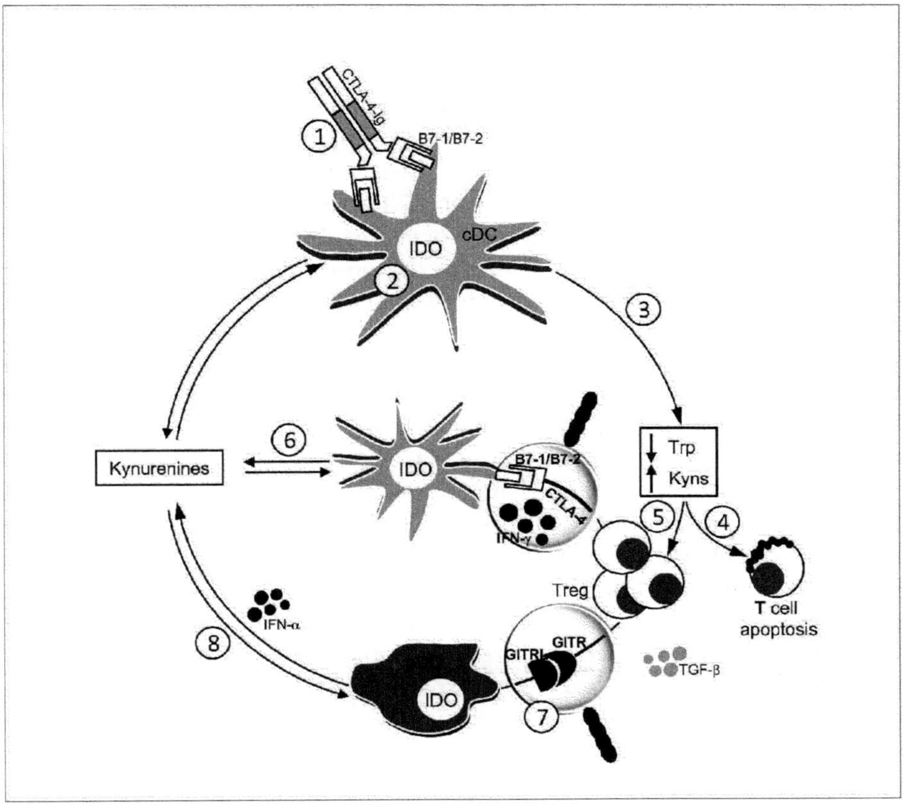

Figure 3
A model of the CTLA-4-Ig/B7 interaction dynamics during an immune response
Ligation of B7-1/B7-2 by CTLA4–Ig on conventional DCs (cDCs) leads to the activation of the enzyme IDO: 1 and 2. Functional expression of IDO in DCs contributes to the activation of tryptophan catabolism in other DC subsets, and the combined effects of tryptophan starvation and tryptophan catabolites creates an immunosuppressive environment that promotes apoptosis of antigen-specific T cells and the generation of new regulatory T cells: 3 through to 6. By means of reverse signaling through B7 molecules and GITRL, Tregs expressing CTLA-4 and/or GITR contribute to additional functional expression of IDO by other cDCs and pDCs; the release of type I and II interferons further potentiate IDO induction: 7 and 8.

For this reason, the CTLA-4-Ig molecules were tested in several animal models of transplantation, such as human pancreatic transplantation in mice, cardiac allograft, renal allografts in rats, bone marrow transplantation and also highly immunogenic skin allografts [103–107]. These studies demonstrated that CTLA-4-Ig was capable of inhibiting T cell-dependent antibody responses and prolong transplanted organ survival [108, 109]. However, CTLA-4-Ig was found to be inadequate to maintain a hyporesponsive state to allograft in some models [110]. The combination of CTLA-4-Ig with anti-CD40L mAb was more potent than either drug alone [111].

Based on these results in preclinical studies, recently a modified form of CTLA-4-Ig LEA29Y (Belatacept) has been generated, which was rationally designed by a mutagenesis strategy to provide higher B7-binding properties than the original CTLA-4-Ig (Abatacept), and therefore probably endowed with even greater immunosuppressive properties [112]. Belatacept demonstrated a fourfold slower off-rate for B7-2 and a twofold slower off-rate for B7-1 compared with Abatacept, and also more potent function [110, 113]. Administration of the new agent to non-human primates, combined with rapamycin and anti-IL-2R mAb resulted in prolonged allograft survival in the absence of concomitant steroids [114]. Whether the mutant Belatacept fusion protein retains the ability or has greater ability to induce IDO in B7-expressing cells has yet to be determined.

Two sets of clinical trials using Abatacept have now been completed, the first one in patients with psoriasis vulgaris, an autoimmune skin disorder and the second one in patients with rheumatoid arthritis (RA). In both a beneficial effect was demonstrated [115, 116]. Based on these trials, Abatacept is now approved for the treatment of patients with RA with an inadequate response to treatment with TNF-α antagonists. Several clinical and preclinical studies suggest a deregulation of the IDO pathway in RA, and a recent one also demonstrates that IDO blockade aggravated the severity of arthritis in a mouse model, suggesting a possible involvement of IDO in the in the pathogenesis of RA.

Regarding the application of CTLA-4-Ig in transplantation, since preclinical studies demonstrated inadequacy of the parental molecule to maintain a hyporesponsive state to the allograft in some models, a Phase III multicenter clinical study is undergoing in primary renal transplant using different regimens of Belatacept in addition to other immunosuppressive drugs [112]. The differential use of Abatacept or Belatacept in RA or transplantation is based on studies in humans or non-human primates showing different requirements for modulating aberrant immune response in autoimmune diseases and blocking the immune response to an allograft. Indeed, several clinical and preclinical studies have reported that IDO induction by CTLA-4 plays an important role. Because of the crucial involvement of IDO activation in CTLA-4-Ig-mediated effects, it is possible that molecules that potentiate and mimic this pathway may synergize with the administration of different forms of CTLA-4-Ig in the treatment of autoimmune diseases or in preventing graft rejection.

References

1 Goodnow CC, Sprent J, Fazekas de St Groth B, Vinuesa CG (2005) Cellular and genetic mechanisms of self tolerance and autoimmunity. *Nature* 435: 590–597
2 Kyewski B, Derbinski J (2004) Self-representation in the thymus: an extended view. *Nat Rev Immunol* 4: 688–698
3 Mathis D, Benoist C (2004) Back to central tolerance. *Immunity* 20: 509–516
4 Venanzi ES, Benoist C, Mathis D (2004) Good riddance: Thymocyte clonal deletion prevents autoimmunity. *Curr Opin Immunol* 16: 197–202
5 Fontenot JD, Rudensky AY (2005) A well adapted regulatory contrivance: regulatory T cell development and the forkhead family transcription factor Foxp3. *Nat Immunol* 6: 331–337
6 Sakaguchi S (2004) Naturally arising $CD4^+$ regulatory t cells for immunologic self-tolerance and negative control of immune responses. *Annu Rev Immunol* 22: 531–562
7 Sakaguchi S, Setoguchi R, Yagi H, Nomura T (2006) Naturally arising Foxp3-expressing $CD25^+CD4^+$ regulatory T cells in self-tolerance and autoimmune disease. *Curr Top Microbiol Immunol* 305: 51–66
8 Filion MC, Proulx C, Bradley AJ, Devine DV, Sekaly RP, Decary F, Chartrand P (1996) Presence in peripheral blood of healthy individuals of autoreactive T cells to a membrane antigen present on bone marrow-derived cells. *Blood* 88: 2144–2150
9 Paroli M, Barnaba V (2005) Mechanisms of $CD8^+$ T cell peripheral tolerance to our own antigens. *Front Biosci* 10: 1628–1634
10 Saouaf SJ, Brennan PJ, Shen Y, Greene MI (2003) Mechanisms of peripheral immune tolerance: conversion of the immune to the unresponsive phenotype. *Immunol Res* 28: 193–199
11 Mills KH (2004) Regulatory T cells: friend or foe in immunity to infection? *Nat Rev Immunol* 4: 841–855
12 Powrie F, Carlino J, Leach MW, Mauze S, Coffman RL (1996) A critical role for transforming growth factor-beta but not interleukin 4 in the suppression of T helper type 1–mediated colitis by CD45RB(low) $CD4^+$ T cells. *J Exp Med* 183: 2669–2674
13 Gavin MA, Rasmussen JP, Fontenot JD, Vasta V, Manganiello VC, Beavo JA, Rudensky AY (2007) Foxp3–dependent programme of regulatory T-cell differentiation. *Nature* 445: 771–775
14 Fontenot JD, Gavin MA, Rudensky AY (2003) Foxp3 programs the development and function of $CD4^+CD25^+$ regulatory T cells. *Nat Immunol* 4: 330–336
15 Hori S, Takahashi T, Sakaguchi S (2003) Control of autoimmunity by naturally arising regulatory $CD4^+$ T cells. *Adv Immunol* 81: 331–371
16 Bluestone JA, Abbas AK (2003) Natural *versus* adaptive regulatory T cells. *Nat Rev Immunol* 3: 253–257
17 Garba ML, Pilcher CD, Bingham AL, Eron J, Frelinger JA (2002) HIV antigens can induce TGF-beta(1)-producing immunoregulatory $CD8^+$ T cells. *J Immunol* 168: 2247–2254

18 Han G, Shao H, Peng Y, Zhang P, Ke Y, Kaplan HJ, Sun D (2007) Suppressor role of rat $CD8^{+}CD45RC^{low}$ T cells in experimental autoimmune uveitis (EAU). *J Neuroimmunol* 183: 81–88

19 Sonoda KH, Faunce DE, Taniguchi M, Exley M, Balk S, Stein-Streilein J (2001) NK T cell-derived IL-10 is essential for the differentiation of antigen-specific T regulatory cells in systemic tolerance. *J Immunol* 166: 42–50

20 Maloy KJ, Salaun L, Cahill R, Dougan G, Saunders NJ, Powrie F (2003) $CD4^{+}CD25^{+}$ T(R) cells suppress innate immune pathology through cytokine-dependent mechanisms. *J Exp Med* 197: 111–119

21 Houot R, Perrot I, Garcia E, Durand I, Lebecque S (2006) Human $CD4^{+}CD25^{high}$ regulatory T cells modulate myeloid but not plasmacytoid dendritic cells activation. *J Immunol* 176: 5293–5298

22 Read S, Greenwald R, Izcue A, Robinson N, Mandelbrot D, Francisco L, Sharpe AH, Powrie F (2006) Blockade of CTLA-4 on $CD4^{+}CD25^{+}$ regulatory T cells abrogates their function *in vivo*. *J Immunol* 177: 4376–4383

23 Nakamura K, Kitani A, Strober W (2001) Cell contact-dependent immunosuppression by CD4(+)CD25(+) regulatory T cells is mediated by cell surface-bound transforming growth factor beta. *J Exp Med* 194: 629–644

24 Belkaid Y, Rouse BT (2005) Natural regulatory T cells in infectious disease. *Nat Immunol* 6: 353–360

25 Watanabe N, Wang YH, Lee HK, Ito T, Wang YH, Cao W, Liu YJ (2005) Hassall's corpuscles instruct dendritic cells to induce $CD4^{+}CD25^{+}$ regulatory T cells in human thymus. *Nature* 436: 1181–1185

26 Steinman RM, Hawiger D, Nussenzweig MC (2003) Tolerogenic dendritic cells. *Annu Rev Immunol* 21: 685–711

27 Gad M, Claesson MH, Pedersen AE (2003) Dendritic cells in peripheral tolerance and immunity. *APMIS* 111: 766–775

28 Tan JK, O'Neill HC (2005) Maturation requirements for dendritic cells in T cell stimulation leading to tolerance *versus* immunity. *J Leukoc Biol* 78: 319–324

29 Lanzavecchia A, Sallusto F (2001) The instructive role of dendritic cells on T cell responses: lineages, plasticity and kinetics. *Curr Opin Immunol* 13: 291–298

30 Mosmann TR, Livingstone AM (2004) Dendritic cells: the immune information management experts. *Nat Immunol* 5: 564–566

31 Fallarino F, Grohmann U, Hwang KW, Orabona C, Vacca C, Bianchi R, Belladonna ML, Fioretti MC, Alegre ML, Puccetti P (2003) Modulation of tryptophan catabolism by regulatory T cells. *Nat Immunol* 4: 1206–1212

32 Mellor AL, Munn DH (2004) IDO expression by dendritic cells: tolerance and tryptophan catabolism. *Nat Rev Immunol* 4: 762–774

33 Grohmann U, Fallarino F, Puccetti P (2003) Tolerance, DCs and tryptophan: much ado about IDO. *Trends Immunol* 24: 242–248

34 Mellor AL, Munn DH (1999) Tryptophan catabolism and T-cell tolerance: immunosuppression by starvation? *Immunol Today* 20: 469–473

35 Sugimoto H, Oda S, Otsuki T, Hino T, Yoshida T, Shir Y (2006) Crystal structure of human indoleamine 2,3-dioxygenase: catalytic mechanism of O2 incorporation by a heme-containing dioxygenase. *Proc Natl Acad Sci USA* 103: 2611–2616

36 Munn DH, Zhou M, Attwood JT, Bondarev I, Conway SJ, Marshall B, Brown C, Mellor AL (1998) Prevention of allogeneic fetal rejection by tryptophan catabolism. *Science* 281: 1191–1193

37 Shortman K, Heath WR (2001) Immunity or tolerance? That is the question for dendritic cells. *Nat Immunol* 2: 988–989

38 Fallarino F, Gizzi S, Mosci P, Grohmann U, Puccetti P (2007) Tryptophan catabolism in IDO^+ plasmacytoid dendritic cells. *Curr Drug Metab* 8: 209–216

39 Braun D, Longman RS, Albert ML (2005) A two-step induction of indoleamine 2,3 dioxygenase (IDO) activity during dendritic-cell maturation. *Blood* 106: 2375–2381

40 Orabona C, Puccetti P, Vacca C, Bicciato S, Luchini A, Fallarino F, Bianchi R, Velardi E, Perruccio K, Velardi A et al (2006) Toward the identification of a tolerogenic signature in IDO-competent dendritic cells. *Blood* 107: 2846–2854

41 Grohmann U, Bianchi R, Belladonna ML, Silla S, Fallarino F, Fioretti MC, Puccetti P (2000) IFN-gamma inhibits presentation of a tumor/self peptide by CD8 $alpha^-$ dendritic cells *via* potentiation of the CD8 $alpha^+$ subset. *J Immunol* 165: 1357–1363

42 Vacca C, Fallarino F, Perruccio K, Orabona C, Bianchi R, Gizzi S, Velardi A, Fioretti MC, Puccetti P, Grohmann U (2005) CD40 ligation prevents onset of tolerogenic properties in human dendritic cells treated with CTLA-4-Ig. *Microbes Infect* 7: 1040–1048

43 Alegre ML, Fallarino F (2006) Mechanisms of CTLA-4–Ig in tolerance induction. *Curr Pharm Des* 12: 149–160

44 Fujigaki H, Saito K, Fujigaki S, Takemura M, Sudo K, Ishiguro H, Seishima M (2006) The signal transducer and activator of transcription 1alpha and interferon regulatory factor 1 are not essential for the induction of indoleamine 2,3-dioxygenase by lipopolysaccharide: involvement of p38 mitogen-activated protein kinase and nuclear factor-kappaB pathways, and synergistic effect of several proinflammatory cytokines. *J Biochem (Tokyo)* 139: 655–662

45 Romani L, Bistoni F, Perruccio K, Montagnoli C, Gaziano R, Bozza S, Bonifazi P, Bistoni G, Rasi G, Velardi A et al (2006) Thymosin alpha1 activates dendritic cell tryptophan catabolism and establishes a regulatory environment for balance of inflammation and tolerance. *Blood* 108: 2265–2274

46 Fallarino F, Asselin-Paturel C, Vacca C, Bianchi R, Gizzi S, Fioretti MC, Trinchieri G, Grohmann U, Puccetti P (2004) Murine plasmacytoid dendritic cells initiate the immunosuppressive pathway of tryptophan catabolism in response to CD200 receptor engagement. *J Immunol* 173: 3748–3754

47 Orabona C, Grohmann U, Belladonna ML, Fallarino F, Vacca C, Bianchi R, Bozza S, Volpi C, Salomon BL, Fioretti MC et al (2004) CD28 induces immunostimulatory signals in dendritic cells *via* CD80 and CD86. *Nat Immunol* 5: 1134–1142

48 Fallarino F, Orabona C, Vacca C, Bianchi R, Gizzi S, Asselin-Paturel C, Fioretti MC, Trinchieri G, Grohmann U, Puccetti P (2005) Ligand and cytokine dependence of the

immunosuppressive pathway of tryptophan catabolism in plasmacytoid dendritic cells. *Int Immunol* 17: 1429–1438

49 Seo SK, Choi JH, Kim YH, Kang WJ, Park HY, Suh JH, Choi BK, Vinay DS, Kwon BS (2004) 4-1BB-mediated immunotherapy of rheumatoid arthritis. *Nat Med* 10: 1088–1094

50 Grohmann U, Volpi C, Fallarino F, Bozza S, Bianchi R, Vacca C, Orabona C, Belladonna ML, Ayroldi E, Nocentini G et al (2007) Reverse signaling through GITR ligand enables dexamethasone to activate IDO in allergy. *Nat Med* 13: 579–586

51 Frumento G, Rotondo R, Tonetti M, Damonte G, Benatti U, Ferrara GB (2002) Tryptophan-derived catabolites are responsible for inhibition of T and natural killer cell proliferation induced by indoleamine 2,3-dioxygenase. *J Exp Med* 196: 459–468

52 Li Y, Tredget EE, Ghaffari A, Lin X, Kilani RT, Ghahary A (2006) Local expression of indoleamine 2,3–dioxygenase protects engraftment of xenogeneic skin substitute. *J Invest Dermatol* 126: 128–136

53 Platten M, Ho PP, Youssef S, Fontoura P, Garren H, Hur EM, Gupta R, Lee LY, Kidd BA, Robinson WH et al (2005) Treatment of autoimmune neuroinflammation with a synthetic tryptophan metabolite. *Science* 310: 850–855

54 Della Chiesa M, Carlomagno S, Frumento G, Balsamo M, Cantoni C, Conte R, Moretta L, Moretta A, Vitale M (2006) The tryptophan catabolite L-kynurenine inhibits the surface expression of NKp46 and NKG2D activating receptors and regulates NK cell function. *Blood* 108: 4118–4125

55 Fallarino F, Grohmann U, Vacca C, Orabona C, Spreca A, Fioretti MC, Puccetti P (2003) T cell apoptosis by kynurenines. *Adv Exp Med Biol* 527: 183–190

56 Fallarino F, Puccetti P (2006) Toll-like receptor 9-mediated induction of the immunosuppressive pathway of tryptophan catabolism. *Eur J Immunol* 36: 8–11

57 Brusko TM, Wasserfall CH, Agarwal A, Kapturczak MH, Atkinson MA (2005) An integral role for heme oxygenase-1 and carbon monoxide in maintaining peripheral tolerance by $CD4^{+}CD25^{+}$ regulatory T cells. *J Immunol* 174: 5181–5186

58 Munn DH, Sharma MD, Baban B, Harding HP, Zhang Y, Ron D, Mellor AL (2005) GCN2 kinase in T cells mediates proliferative arrest and anergy induction in response to indoleamine 2,3-dioxygenase. *Immunity* 22: 633–642

59 Bretscher PA (1999) A two-step, two-signal model for the primary activation of precursor helper T cells. *Proc Natl Acad Sci USA* 96: 185–190

60 Van Parijs L, Abbas AK (1998) Homeostasis and self-tolerance in the immune system: turning lymphocytes off. *Science* 280: 243–248

61 Yamada A, Salama AD, Sayegh MH (2002) The role of novel T cell costimulatory pathways in autoimmunity and transplantation. *J Am Soc Nephrol* 13: 559–575

62 Collins M, Ling V, Carreno BM (2005) The B7 family of immune-regulatory ligands. *Genome Biol* 6: 223

63 McAdam AJ, Schweitzer AN, Sharpe AH (1998) The role of B7 co-stimulation in activation and differentiation of $CD4^{+}$ and $CD8^{+}$ T cells. *Immunol Rev* 165: 231–247

64 Sloan-Lancaster J, Evavold BD, Allen PM (1993) Induction of T-cell anergy by altered T-cell-receptor ligand on live antigen-presenting cells. *Nature* 363: 156–159
65 Linsley PS, Greene JL, Tan P, Bradshaw J, Ledbetter JA, Anasetti C, Damle NK (1992) Coexpression and functional cooperation of CTLA-4 and CD28 on activated T lymphocytes. *J Exp Med* 176: 1595–1604
66 Linsley PS, Greene JL, Brady W, Bajorath J, Ledbetter JA, Peach R (1994) Human B7-1 (CD80) and B7-2 (CD86) bind with similar avidities but distinct kinetics to CD28 and CTLA-4 receptors. *Immunity* 1: 793–801
67 Greene JL, Leytze GM, Emswiler J, Peach R, Bajorath J, Cosand W, Linsley PS (1996) Covalent dimerization of CD28/CTLA-4 and oligomerization of CD80/CD86 regulate T cell costimulatory interactions. *J Biol Chem* 271: 26762–26771
68 Chambers CA, Kuhns MS, Allison JP (1999) Cytotoxic T lymphocyte antigen-4 (CTLA-4) regulates primary and secondary peptide-specific CD4(+) T cell responses. *Proc Natl Acad Sci USA* 96: 8603–8608
69 Read S, Malmstrom V, Powrie F (2000) Cytotoxic T lymphocyte-associated antigen 4 plays an essential role in the function of CD25(+)CD4(+) regulatory cells that control intestinal inflammation. *J Exp Med* 192: 295–302
70 Jago CB, Yates J, Camara NO, Lechler RI, Lombardi G (2004) Differential expression of CTLA-4 among T cell subsets. *Clin Exp Immunol* 136: 463–471
71 Finn PW, He H, Wang Y, Wang Z, Guan G, Listman J, Perkins DL (1997) Synergistic induction of CTLA-4 expression by costimulation with TCR plus CD28 signals mediated by increased transcription and messenger ribonucleic acid stability. *J Immunol* 158: 4074–4081
72 Anjos S, Nguyen A, Ounissi-Benkalha H, Tessier MC, Polychronakos C (2002) A common autoimmunity predisposing signal peptide variant of the cytotoxic T-lymphocyte antigen 4 results in inefficient glycosylation of the susceptibility allele. *J Biol Chem* 277: 46478–46486
73 Sharpe AH, Freeman GJ (2002) The B7–CD28 superfamily. *Nat Rev Immunol* 2: 116–126
74 Stamper CC, Zhang Y, Tobin JF, Erbe DV, Ikemizu S, Davis SJ, Stahl ML, Seehra J, Somers WS, Mosyak L (2001) Crystal structure of the B7-1/CTLA-4 complex that inhibits human immune responses. *Nature* 410: 608–611
75 Schwartz JC, Zhang X, Fedorov AA, Nathenson SG, Almo SC (2001) Structural basis for co-stimulation by the human CTLA-4/B7-2 complex. *Nature* 410: 604–608
76 Baroja ML, Madrenas J (2003) Viewpoint: therapeutic implications of CTLA-4 compartmentalization. *Am J Transplant* 3: 919–926
77 Darlington PJ, Baroja ML, Chau TA, Siu E, Ling V, Carreno BM, Madrenas J (2002) Surface cytotoxic T lymphocyte-associated antigen 4 partitions within lipid rafts and relocates to the immunological synapse under conditions of inhibition of T cell activation. *J Exp Med* 195: 1337–1347
78 Chitnis T, Najafian N, Abdallah KA, Dong V, Yagita H, Sayegh MH, Khoury SJ (2001)

CD28-independent induction of experimental autoimmune encephalomyelitis. *J Clin Invest* 107: 575–583
79 Bachmaier K, Krawczyk C, Kozieradzki I, Kong YY, Sasaki T, Oliveira-dos-Santos A, Mariathasan S, Bouchard D, Wakeham A, Itie A et al (2000) Negative regulation of lymphocyte activation and autoimmunity by the molecular adaptor Cbl-b. *Nature* 403: 211–216
80 Li D, Gal I, Vermes C, Alegre ML, Chong AS, Chen L, Shao Q, Adarichev V, Xu X, Koreny T et al (2004) Cutting edge: Cbl-b: one of the key molecules tuning CD28- and CTLA-4–mediated T cell costimulation. *J Immunol* 173: 7135–7139
81 Vijayakrishnan L, Slavik JM, Illes Z, Rainbow D, Peterson LB, Sharpe AS, Wicker LS, Kuchroo VK (2005) An autoimmune disease-associated CTLA4 splice variant lacking the B7 binding domain signals negatively in T cells. *Novartis Found Symp* 267: 200–212; discussion 212–218
82 Chikuma S, Abbas AK, Bluestone JA (2005) B7–independent inhibition of T cells by CTLA-4. *J Immunol* 175: 177–181
83 Thompson CB, Lindsten T, Ledbetter JA, Kunkel SL, Young HA, Emerson SG, Leiden JM, June CH (1989) CD28 activation pathway regulates the production of multiple T-cell-derived lymphokines/cytokines. *Proc Natl Acad Sci USA* 86: 1333–1337
84 Lindstein T, June CH, Ledbetter JA, Stella G, Thompson CB (1989) Regulation of lymphokine messenger RNA stability by a surface-mediated T cell activation pathway. *Science* 244: 339–343
85 Noel PJ, Boise LH, Green JM, Thompson CB (1996) CD28 costimulation prevents cell death during primary T cell activation. *J Immunol* 157: 636–642
86 Tivol EA, Borriello F, Schweitzer AN, Lynch WP, Bluestone JA, Sharpe AH (1995) Loss of CTLA-4 leads to massive lymphoproliferation and fatal multiorgan tissue destruction, revealing a critical negative regulatory role of CTLA-4. *Immunity* 3: 541–547
87 Waterhouse P, Penninger JM, Timms E, Wakeham A, Shahinian A, Lee KP, Thompson CB, Griesser H, Mak TW (1995) Lymphoproliferative disorders with early lethality in mice deficient in Ctla-4. *Science* 270: 985–988
88 Brunner MC, Chambers CA, Chan FK, Hanke J, Winoto A, Allison JP (1999) CTLA-4-mediated inhibition of early events of T cell proliferation. *J Immunol* 162: 5813–5820
89 Allison JP, Krummel MF (1995) The Yin and Yang of T cell costimulation. *Science* 270: 932–933
90 Takahashi S, Kataoka H, Hara S, Yokosuka T, Takase K, Yamasaki S, Kobayashi W, Saito Y, Saito T (2005) *In vivo* overexpression of CTLA-4 suppresses lymphoproliferative diseases and thymic negative selection. *Eur J Immunol* 35: 399–407
91 Mandelbrot DA, McAdam AJ, Sharpe AH (1999) B7–1 or B7–2 is required to produce the lymphoproliferative phenotype in mice lacking cytotoxic T lymphocyte-associated antigen 4 (CTLA-4). *J Exp Med* 189: 435–440
92 Tivol EA, Boyd SD, McKeon S, Borriello F, Nickerson P, Strom TB, Sharpe AH (1997) CTLA4Ig prevents lymphoproliferation and fatal multiorgan tissue destruction in CTLA-4–deficient mice. *J Immunol* 158: 5091–5094

93 Carreno BM, Bennett F, Chau TA, Ling V, Luxenberg D, Jussif J, Baroja ML, Madrenas J (2000) CTLA-4 (CD152) can inhibit T cell activation by two different mechanisms depending on its level of cell surface expression. *J Immunol* 165: 1352–1356

94 Lin H, Rathmell JC, Gray GS, Thompson CB, Leiden JM, Alegre ML (1998) Cytotoxic T lymphocyte antigen 4 (CTLA4) blockade accelerates the acute rejection of cardiac allografts in CD28-deficient mice: CTLA4 can function independently of CD28. *J Exp Med* 188: 199–204

95 Grohmann U, Orabona C, Fallarino F, Vacca C, Calcinaro F, Falorni A, Candeloro P, Belladonna ML, Bianchi R, Fioretti MC et al (2002) CTLA-4-Ig regulates tryptophan catabolism *in vivo*. *Nat Immunol* 3: 1097–1101

96 Dunn GP, Bruce AT, Ikeda H, Old LJ, Schreiber RD (2002) Cancer immunoediting: from immunosurveillance to tumor escape. *Nat Immunol* 3: 991–998

97 Pardoll D (2003) Does the immune system see tumors as foreign or self? *Annu Rev Immunol* 21: 807–839

98 Widner B, Sepp N, Kowald E, Ortner U, Wirleitner B, Fritsch P, Baier-Bitterlich G, Fuchs D (2000) Enhanced tryptophan degradation in systemic lupus erythematosus. *Immunobiology* 201: 621–630

99 Forrest CM, Kennedy A, Stone TW, Stoy N, Darlington LG (2003) Kynurenine and neopterin levels in patients with rheumatoid arthritis and osteoporosis during drug treatment. *Adv Exp Med Biol* 527: 287–295

100 Boasso A, Herbeuval JP, Hardy AW, Anderson SA, Dolan MJ, Fuchs D, Shearer GM (2006) HIV-1 inhibits $CD4^+$ T cell proliferation by inducing indoleamine 2,3-dioxygenase in plasmacytoid dendritic cells. *Blood* 109: 3351–3359

101 Roncarolo MG, Bacchetta R, Bordignon C, Narula S, Levings MK (2001) Type 1 T regulatory cells. *Immunol Rev* 182: 68–79

102 Serafini P, De Santo C, Marigo I, Cingarlini S, Dolcetti L, Gallina G, Zanovello P, Bronte V (2004) Derangement of immune responses by myeloid suppressor cells. *Cancer Immunol Immunother* 53: 64–72

103 Lenschow DJ, Zeng Y, Thistlethwaite JR, Montag A, Brady W, Gibson MG, Linsley PS, Bluestone JA (1992) Long-term survival of xenogeneic pancreatic islet grafts induced by CTLA4lg. *Science* 257: 789–792

104 Turka LA, Linsley PS, Lin H, Brady W, Leiden JM, Wei RQ, Gibson ML, Zheng XG, Myrdal S, Gordon D et al (1992) T-cell activation by the CD28 ligand B7 is required for cardiac allograft rejection *in vivo*. *Proc Natl Acad Sci USA* 89: 11102–11105

105 Azuma H, Chandraker A, Nadeau K, Hancock WW, Carpenter CB, Tilney NL, Sayegh MH (1996) Blockade of T-cell costimulation prevents development of experimental chronic renal allograft rejection. *Proc Natl Acad Sci USA* 93: 12439–12444

106 Wekerle T, Kurtz J, Ito H, Ronquillo JV, Dong V, Zhao G, Shaffer J, Sayegh MH, Sykes M (2000) Allogeneic bone marrow transplantation with co-stimulatory blockade induces macrochimerism and tolerance without cytoreductive host treatment. *Nat Med* 6: 464–469

107 Larsen CP, Elwood ET, Alexander DZ, Ritchie SC, Hendrix R, Tucker-Burden C, Cho

HR, Aruffo A, Hollenbaugh D, Linsley PS et al (1996) Long-term acceptance of skin and cardiac allografts after blocking CD40 and CD28 pathways. *Nature* 381: 434–438

108 Russell ME, Hancock WW, Akalin E, Wallace AF, Glysing-Jensen T, Willett TA, Sayegh MH (1996) Chronic cardiac rejection in the LEW to F344 rat model. Blockade of CD28–B7 costimulation by CTLA4Ig modulates T cell and macrophage activation and attenuates arteriosclerosis. *J Clin Invest* 97: 833–838

109 Chandraker A, Azuma H, Nadeau K, Carpenter CB, Tilney NL, Hancock WW, Sayegh MH (1998) Late blockade of T cell costimulation interrupts progression of experimental chronic allograft rejection. *J Clin Invest* 101: 2309–2318

110 Levisetti MG, Padrid PA, Szot GL, Mittal N, Meehan SM, Wardrip CL, Gray GS, Bruce DS, Thistlethwaite JR Jr, Bluestone JA (1997) Immunosuppressive effects of human CTLA4Ig in a non-human primate model of allogeneic pancreatic islet transplantation. *J Immunol* 159: 5187–5191

111 Kirk AD, Harlan DM, Armstrong NN, Davis TA, Dong Y, Gray GS, Hong X, Thomas D, Fechner JH Jr, Knechtle SJ (1997) CTLA4-Ig and anti-CD40 ligand prevent renal allograft rejection in primates. *Proc Natl Acad Sci USA* 94: 8789–8794

112 Vincenti F (2005) Chronic induction. What's new in the pipeline. *Contrib Nephrol* 146: 22–29

113 Larsen CP, Pearson TC, Adams AB, Tso P, Shirasugi N, Strobertm E, Anderson D, Cowan S, Price K, Naemura J et al (2005) Rational development of LEA29Y (belatacept), a high-affinity variant of CTLA4–Ig with potent immunosuppressive properties. *Am J Transplant* 5: 443–453

114 Adams AB, Shirasugi N, Durham MM, Strobert E, Anderson D, Rees P, Cowan S, Xu H, Blinder Y, Cheung M et al (2002) Calcineurin inhibitor-free CD28 blockade-based protocol protects allogeneic islets in nonhuman primates. *Diabetes* 51: 265–270

115 Abrams JR, Lebwohl MG, Guzzo CA, Jegasothy BV, Goldfarb MT, Goffe BS, Menter A, Lowe NJ, Krueger G, Brown MJ et al (1999) CTLA4Ig-mediated blockade of T-cell costimulation in patients with psoriasis vulgaris. *J Clin Invest* 103: 1243–1252

116 Weinblatt M, Combe B, Covucci A, Aranda R, Becker JC, Keystone E (2006) Safety of the selective costimulation modulator abatacept in rheumatoid arthritis patients receiving background biologic and nonbiologic disease-modifying antirheumatic drugs: A one-year randomized, placebo-controlled study. *Arthritis Rheum* 54: 2807–2816

Adhesion molecules as therapeutic targets

Mark R. Nicolls and Rasa Tamosiuniene

Veterans Administration Palo Alto, Stanford University, Departments of Medicine and Division of Pulmonary and Critical Care Medicine, 3801 Miranda Ave., Palo Alto, CA 94304, USA

Immunoglobulin supergene family

The immunoglobulin supergene family (IgSF) cell adhesion molecules (CAMs) are either homophilic or heterophilic proteins that bind either integrins or different IgSF CAMs. Proteins are classified into the IgSF if they possess a structural domain known as an Ig domain, which contain about 70–110 amino acids and are categorized into different types according to their size and function [1]. Members of this family with important adhesion function include CD2, CD48, the SIGLEC family (sialic acid binding Ig-like lectins such as CD22, CD83), intracellular adhesion molecules (ICAMs), vascular cell adhesion molecule (VCAM-1), platelet-endothelial cell adhesion molecule (PECAM-1), neural cell adhesion molecules (NCAMs), L1-CAM and CHL-1. NCAMs, L1-CAM and CHL-1 are important proteins for neurological cell adhesion. Members of the IgSF are ligands for the integrins, and so these can be considered together conceptually with this class of CAMs. IgSF members have not been as extensively targeted as have integrin proteins, and the broad utility of these agents are largely undetermined. A summary of selected pre-clinical studies and clinical trials is presented in Table 1. Anti-ICAM therapy has proven neutral or possibly deleterious in a stroke trial [2]. Anti-CD2 therapy has shown efficacy against graft vs. host disease [3], but this is likely due to lymphocyte and natural killer cell depletion rather than anti-adhesion effects.

Integrins

Integrin CAMs (particularly leukocyte function associated antigen-1, LFA-1) are by far the most extensively studied anti-adhesion targets. Integrins are heterodimeric receptors composed of noncovalently linked α- and β-subunits [4]. Leukocytes can express at least 12 of the 24 known integrin heterodimers depending on the maturation of the cell. Integrins are critically involved in the trafficking of leukocytes into tissues. The most important integrins for leukocyte migration are members of the β_2 integrin family, especially $\alpha_L\beta_2$ (CD11a/CD18 or LFA-1), $\alpha_M\beta_2$ (CD11b/CD18

The Immune Synapse as a Novel Target for Therapy, edited by Luis Graca

Table 1 - Selected preclinical and clinical studies targeting IgSF members.

IgSF Member	Condition treated	Therapy	Treatment outcome	Stage of development	Ref.
ICAM-1	Stroke	Murine mAb against ICAM-1 (Enlimomab)	Stroke patients with worsened clinical outcomes	Phase III (stopped)	[2]
	Rheumatoid arthritis	Murine mAb against ICAM-1	Clinical improvement in early or indolent rheumatoid arthritis with first dose of therapy	Phase I/II	[108]
	Renal transplantation	Murine mAb against ICAM-1 (Enlimomab)	Induction therapy not effective in reducing the risk of delayed onset of graft function in cadaveric renal transplant recipients	Phase III	[109]
VCAM-1	Stroke	Murine anti-VCAM-1	Blocking VCAM-1 not effective in stroke reduction	Pre-clinical	[110]
PECAM-1	Myocardial ischemia-reperfusion (I-R) injury	F(ab')$_2$ fragments of anti-PECAM-1 antibody	Blocking PECAM-1 significantly reduces myocardial I-R injury in rats and cats	Pre-clinical	[111, 112]

or Mac-1) and the two α_4 integrins, $\alpha_4\beta_1$ (VLA-4) and $\alpha_4\beta_7$. Leukocyte integrins bind to endothelial ligands which are members of the IgSF and include: ICAM-1, ICAM-2, VCAM-1, PECAM-1, mucosal adressin cell adhesion molecule-1 (MAdCAM-1), junctional adhesion molecule (JAM), and receptor for advance glycation end products (RAGE) [4]. Table 2 outlines pre-clinical studies using integrin antagonists and Table 3 describes clinical trials targeting integrin adhesion molecules. More attention has been historically focused on LFA-1 than on any other integrin target [5], although more recent high impact studies have centered on α_4 integrin antagonists.

LFA-1 was one of earliest of cell-surface molecules identified by monoclonal antibodies (mAbs) generated against leukocyte immunogens and has been perhaps the most thoroughly studied adhesion therapy to date. While there has been long-standing interest in LFA-1 as a therapeutic target for regulating immunity, anti-LFA-

Table 2 - Selected pre-clinical models using integrin antagonists.

Integrin	Condition treated	Therapy added	Treatment outcome	Ref.
CD11a (LFA-1)	Heart transplantation	None, anti-ICAM-1, cyclosporine, BMT + Anti-CD40L, BMT + Everolimus	Prolongs allograft transplantation survival and promotes immune tolerance	[25, 56, 58–60, 113]
	Islet transplantation	None, anti-CD40L, anti-CD45RB	Prolongs allograft and xenograft survival and promotes immune tolerance. Combinational strategies best	[27, 47, 48, 77]
	Tracheal transplantation	None, anti-CD40L	Prolongs allograft survival	[62, 63]
	Hepatocellular transplantation	None, anti-CD40L	Prolongs allograft survival	[79]
	Asthma	None	Prevents airway hyperreactivity	[114]
$\alpha_4\beta_4$ integrin (VLA-4)	Multiple sclerosis (EAE)	None	Protective or exacerbating depending on timing of administration	[115, 116]
$\alpha_4\beta_7$ integrin	Experimental inflammatory bowel disease	None	Inhibits lymphocyte homing	[117]

1 therapy is still not a first-line indication for any clinical condition. Antagonism of LFA-1 with mAbs, either alone or in combination with other agents, can result in regulatory tolerance *in vivo*. Furthermore, new generation humanized anti-LFA-1 monoclonal antibodies (efalizumab) show at least modest promise for continued application in clinical trials. Thus, anti-LFA-1 forms a potential, but still largely unexploited, immunotherapy that may find its greatest application as an agent that augments other therapies.

LFA-1 was first identified in mice in 1981 [6] and in humans in 1982 [7]. Blockade of this molecule notably inhibited target cell killing by cytotoxic T cells. LFA-1 interacts primarily with ICAM-1, but also forms distinct ligands with ICAM-2, ICAM-3 as well as JAM-1 [8]. Early studies suggested that the primary role of LFA-1 was to enhance cell-cell interactions between leukocytes. This function forms the basis of the well-appreciated role of LFA-1 as an adhesion molecule involved in

Table 3 - Selected clinical trials using integrin antagonists.

Integrin	Condition treated	Agent	Trial (highest phase)	Treatment outcome	Ref.
CD11a (LFA-1)	Renal transplantation	Efalizumab	Phase II	Treatment added to one of two immunosuppression regimens. High rate of post-transplant lymphoproliferative disease in patients also receiving full-dose cyclosporine	[89]
	Bone marrow transplantation	Mouse-anti-LFA-1	Phase II	Efficacious in pediatric but not in adult leukemia patients	[83, 84]
	Asthma	Efalizumab	Phase III	Reduced allergen-induced airway inflammation but little impact on airflows	[92]
	Psoriasis	Efalizumab	Phase III	Diminishes disease and improves quality of life	[94]
CD18/ CD11a	Myocardial infarction	Rovelizumab	Phase III (stopped)	No reduction in infarct size in patients undergoing primary angioplasty	[118]
α_4 Integrins	Multiple sclerosis	Natalizumab (targets $\alpha_4\beta_7$, $\alpha_4\beta_1$)	Phase III (stopped)	Two phase III trials demonstrating efficacy with the second study being terminated because of the occurrence of progressive multifocal leukoencelphalopathy (PML) in two patients receiving this Natalizumab in addition to interferon beta	[99, 100]
	Crohn's disease	Natalizumab	Phase III	Effective therapy for Crohn's disease in several large trials. Three patients in the ENACT studies develop PML	[119–121]
	Asthma	GW559090X $\alpha_4\beta_1$ (VLA-4) antagonist	Phase III	Not effective	[122]

leukocyte binding and trafficking [9]. Indeed, LFA-1 greatly increases the functional avidity of T cell–antigen-presenting cell (APC) interactions [10]. As such, LFA-1 appeared to be an attractive target for clinical disease.

The integrins LFA-1 and its ligand ICAM play a critical role in the architecture of the immunological synapse. For antigen-specific recognition, the immunological synapse can be generally defined as the physical structure of the interacting surfaces of T cells and APCs, and generally consists of an external ring of LFA-1 surrounding a central TCR-rich area [11–14]. At the beginning of T cell activation, LFA-1 is centered at the T cell–APC contact region with the TCR localizing mainly at the periphery of the synapse [11, 12, 15]. As LFA-1 rapidly clusters following T cell engagement, it is likely that LFA-1-mediated adhesion optimizes T cell–APC contact and increases the number of engaged TCRs. As the immunological synapse matures, the TCR moves to the center of the T cell–APC contact site as LFA-1 moves to the periphery colocalized with the cytoskeletal protein talin [11, 12, 15]. After 1 h of T cell engagement, the TCR is largely absent, while clustered LFA-1 is present for at least 4 h [12, 15, 16]. This sequence of events has been interpreted to mean that LFA-1 is integral to the stabilization of T cell engagement with the APC, and to the optimal activation of T cells [17]. LFA-1 is expressed in higher concentrations in memory T cells compared to naïve T cells [18–20] and likely affects the ability of memory T cells to scan immune targets and form immunological synapses [17]. In summary, LFA-1 appears to be a particularly important integrin in the immunological synapse, and disruption of LFA-1 activity likely strongly affects the stability of this immune interface.

While LFA-1 has been primarily characterized on T cells, it is also present on B cells and may play a role in antigen presentation. Ligation of LFA-1 by immobilized ICAM-1 increases the efficiency of B cell antigen presentation [21]. Human PBMCs responses to B cell mitogens are inhibited by anti-LFA-1, but this effect is likely due to an indirect effect of anti-LFA-1 on monocytes and/or T cells rather than on B cells [22, 23]. Anti-LFA-1 added to culture strongly inhibits antibody responses, although B cell proliferative responses are only partially mitigated [24]. Normally, antibody production requires interaction between B cells, T cells and dendritic cells. In culture, these three cell populations cluster together within 1 h, and this clustering is inhibited by anti-LFA-1, suggesting a role for LFA-1 in this initial association. Relatively little is known about how anti-LFA-1 monotherapy affects antibody responses following transplantation, although several studies have demonstrated that in combination with other agents both alloantibody [25] and xenoantibody [26, 27] responses are inhibited. In summary, not only is LFA-1 integral to the structure of the immunological synapse, but it is possible that LFA-1 targeted therapy may also interfere with normal B cell responses.

In any inflammatory response, lymphocytes must transmigrate through the vascular endothelium to arrive at the site of injury. Numerous studies have demonstrated a role for LFA-1 in lymphocyte transendothelial migration [28], and anti-LFA-1 has a demonstrable impact on lymphocyte trafficking from peripheral blood to lymph nodes [29]. Rolling lymphocytes on the vascular endothelium are arrested prior to diapedesis into tissues, and LFA-1 is the dominant integrin involved in lymphocyte arrest [30]. In this situation, it is endothelium-displayed chemokines that activates LFA-1 through G protein-coupled receptors [30, 31]. By blocking LFA-1, it is possible that endothelial changes in the vicinity of inflammation have less capacity to stimulate lymphocyte recruitment across the endothelial surface. In LFA-1-deficient mice, both neutrophils and activated T cells were unable to cross endothelial cell monolayers in response to a chemokine gradient, which contrasted with the behavior of wild-type T cells [32]. It is possible that the non-lymphocyte-depleting properties of anti-LFA-1 relate to interference with migration from the vascular space. Given a potential pathogenic role for neutrophils in early rejection events [33], a potential, but as yet unproven, additional benefit of anti-LFA-1 therapy could therefore be to limit the early infiltration of neutrophils. Anti-LFA-1 therapy down modulates the LFA-1 molecule *in vivo* [34] and has already been shown to prevent transmigration of pathogenic lymphocytes implicated in destructive immunity [35], which again points to the potential of anti-LFA-1 to act by temporarily interfering with the homing activity of a primed immune response.

There have been studies suggesting that LFA-1 can provide important costimulatory signals to resting T cells [36, 37] even in the absence of other known costimulatory molecules [38]. However, in contrast to the extensive data available concerning signals involved in TCR, CD40L and CD28 ligation, much less is known about specific signals generated by adhesion-mediated events in the immunological synapse. Studies have shown that increasing antigen density by more than 10 000-fold does not initiate naïve $CD4^+$ T cell proliferation or cytokine synthesis in the absence of LFA-1–ICAM-1 interaction [39, 40]. The β2 subunit of LFA-1 has been implicated as important for signaling events thought to be associated with this LFA-1–ICAM-1 engagement [41, 42]. LFA-1–ICAM-1 interactions may provide stronger TCR–APC adhesion that facilitates more pronounced signaling through other ligand-receptor pairs. Despite indications that LFA-1–ICAM-1 interactions can lead to sustained intracellular calcium levels, elevated inositol phospholipid hydrolysis and the appearance of the hyperphosphorylated p23 form of the TCRζ chain [40, 43, 44], the functional effects of an LFA-1-mediated costimulatory signals are not well defined, although recent work has focused on signaling of kinase Erk1/2 through cytohesin-1 [45, 46].

Therapies that antagonize LFA-1 may exert its effects by blocking some or all of these unique aforementioned signaling pathways critical to the formation of the immunological synapse and to LFA-1-mediated costimulation. While it is clearly possible that anti-LFA-1 also exerts effects by impeding early trafficking of lymphocytes to a site of injury or inflammation, we have observed that accepted islet allografts in animals that have received anti-LFA-1 monotherapy ultimately develop significant lymphocytic infiltrates [47]. If anti-LFA-1 worked only by inducing immune ignorance by blocking lymphocyte trafficking to the graft, it cannot explain the ability of these animals to resist rejection of their transplants following immunization with donor-type spleen cells [47, 48] or the ability to adoptively transfer specific tolerance. On the level of intercellular interactions, the interaction between LFA-1 and the corresponding ICAM-1 ligand promotes Th1 immunity; and inhibiting this interaction can result in increased Th2 cytokine (IL-4 and IL-5) production by activated T cells [49–51]. One hypothesis is that by blocking LFA-1, signal 1 is similarly blocked and that this is the basis for therapeutic effect (rather than blockade of a unique costimulatory pathway) [10]. This hypothesis has been disputed [49] for the following reasons: (1) ICAM antagonism had a markedly different effect than reducing the antigen (i.e., signal 1) concentration; (2) the increase in Th2 cytokines was much more dramatic by blocking LFA-1–ICAM-1 interaction than it was by lowering the antigen dose; (3) regardless of the antigen concentration, blocking LFA-1/ICAM interacted always resulted in an increase in Th2 cytokines; and finally (4) lower antigen doses always resulted in lower IFN-γ concentration, whereas blocking LFA-1–ICAM interactions had no such effect. The varied functions of LFA-1 make it difficult to predict the exact consequence of therapies targeting this molecule. Blockade of LFA-1 may result in inhibition of leukocyte trafficking, altered T and B cell function, or a combination of these effects.

Antibodies interfering with LFA-1–ICAM-1 interactions have been extensively evaluated in numerous pre-clinical studies, which have shown variable efficacy in a variety of solid organ and cellular transplants [25, 27, 47, 48, 52–80]. In these studies, interference with LFA-1 is often combined with additional therapy to enhance immunosuppressive effects. Anti-LFA-1 is also one of a number of agents to prevent autoimmune diabetes in NOD mice [81, 82]. As Table 2 shows, many studies have examined the use of anti-LFA-1 given with other therapies, which likely reflects the inability of anti-LFA-1 monotherapy to provide adequate coverage in stringent model systems. Among combinational strategies, anti-LFA-1 has been used synergistically with anti-ICAM-1, calcineurin inhibition, everolimus, and anti-CD40L.

The clinical application of LFA-1 antagonism began in the 1990s and has yet to establish a first-line indication for anti-LFA-1. Combined anti-LFA-1 and anti-CD2 mAb therapy proved highly effective in bone marrow transplantation for high-risk acute lymphoblastic leukemia in children [83]. However, the early success of anti-LFA-1 therapy in preventing bone marrow transplant rejection in children was generally not observed in adults [84]. Similarly, the early experience of anti-LFA-1 therapy in kidney transplantation showed mixed results. An initial pilot study to block human kidney allograft rejection with anti-LFA-1 therapy employing a mouse anti-human CD11a antibody failed to show efficacy [85]. A subsequent larger multicenter study with this same mAb established that anti-LFA-1 treatment achieved the same results as T cell-depleting rabbit anti-thymocyte globulin treatment for induction therapy in renal transplantation [86]. Given the current uncertainty about the efficacy of T cell-depletion induction strategies [87], this study is of unclear significance.

Efalizumab, a humanized IgG1 anti-LFA-1 antibody has been used in several clinical conditions including kidney transplantation and psoriasis with variable efficacy. Efalizumab effectively blocks LFA-1–ICAM-1 interactions and inhibits T cell activation [88]. In a large efficacy and safety trial that enrolled 556 psoriatic patients, efalizumab was associated with a <4% incidence of anti-efalizumab antibodies. In a Phase I/II open label, dose ranging, multicenter trial, efalizumab (either high dose or low dose was administered following renal transplantation [89]. Although well tolerated, a small subset of patients (3/10 patients) receiving high-dose efalizumab developed post-transplant lymphoproliferative disease. The latter result was highly concerning and has suggested heightened caution with this agent, especially in the setting of increased immunosuppression. Future clinical studies may benefit by employing lower dosing regimens to minimize the risks of post-transplant lymphoproliferative disorder or by using it in the setting of less global immunosuppression (i.e., calcineurin inhibitors, glucocorticosteroids) and in combination with more immune-selective therapeutics.

Recently, the effects of LFA-1 antagonism have been examined in allergen-induced airway responses given that LFA-1 is an important adhesion molecule involved in the migration of neutrophils and lymphocytes to the lung [90, 91]. A recent clinical study demonstrates that eosinophils, mast cells and basophils are also influenced by LFA-1 trafficking [92]. In this randomized, double-blinded, placebo-controlled, parallel group, multicenter study, efalizumab was found to have little impact on airflows compared to placebo but reduced the post allergen increase in activated eosinophils, mast cells and basophils. The authors of this study noted that LFA-1 antagonism appear to inhibit allergen-induced cellular inflammatory responses and could potentially attenuate late asthmatic responses. In a similar vein, LFA-1 expression on $CD4^+$ cells is increased in bronchoalveolar lavage fluid of chronic berylliosis patients, and *in vitro* proliferative responses and cytokine release is markedly reduced by anti-LFA-1 [93]. In this setting, anti-LFA-1 has been conjectured as a therapy for chronic beryllium disease. In summary, it is possible that there may be a role LFA-1 targeting in certain immune-mediated lung conditions, but this application requires much greater study before broader conclusions can be reached.

While still not a first-line therapy for psoriasis, anti-LFA-1 therapy (without additional immunosuppression) is of unequivocal benefit in the treatment of this highly prevalent, and immunologically mediated condition. Efalizumab has been used in phase III trials in patients with psoriasis (reviewed in [94]) that document improved quality of life and diminished disease. Common adverse event associated with anti-LFA-1 therapy in these studies included headache, chills, fever, nausea, vomiting and myalgias. It should be emphasized that, although few direct comparisons are available, efalizumab is substantially less effective in psoriasis than established therapies including cyclosporine, methotrexate and ultraviolet radiation-based therapies (oral psoralen UV-A, UV-B) (reviewed in [95]). Nonetheless, the principle is being established for this skin disease that anti-LFA-1 therapy can generally be safely self administered with subcutaneous injections for periods as long as 1–2 years [94]. In conclusion, anti-LFA-1 therapy has slowly found wider clinical application in autoimmune, alloimmune and allergic diseases, but to date, it appears to have no central first-line indications. As appropriate uses for anti-LFA-1 therapy are being delineated, ongoing use of this biological agent in psoriasis will likely provide invaluable information about dosing, safety and *in vivo* activity.

Therapeutic targeting of α_4 integrin, the other main pathway responsible for leukocyte arrest on vascular endothelium has been evaluated in multiple sclerosis, inflammatory bowel disease and asthma. As with LFA-1 blockade, interference with this group of integrins may impact more than leukocyte trafficking. For example, the integrin $\alpha_4\beta_1$ (VLA-4) not only mediates the adhesion but also provides costimulatory signals that contribute to the activation of T cells [96]. It has recently been demonstrated that $\alpha_4\beta_1$ is recruited in both human and mouse antigen-dependent immune synapses, when the APC is a B cell or a dendritic cell. $\alpha_4\beta_1$ colocalizes with LFA-1 at the peripheral supramolecular activation complex. Targeting VLA-4 with anti-α_4 antibodies results in VLA-4 colocalizing with the CD3-ζ chain at the center of the synapse. Additionally, antibody engagement of the α_4 integrin induces an immune deviation to Th1 response that dampens Th2 autoimmune responses [96]. Therefore, like LFA-1-directed therapies, which involve interference with normal adhesion, costimulation and the immune synapse, anti-α4 antibodies may also work through several of these aforesaid mechanisms.

The most significantly considered α4 integrin-antagonizing therapy to date is natalizumab, which consists of humanized neutralizing IgG4κ mAbs against the leukocyte α_4 integrins (including $\alpha_4\beta_7$ and $\alpha_4\beta_1$) found on lymphocytes and monocytes. Natalizumab has been used to treat conditions involving the invasion of lymphocytes and/or eosinophils, and both cell types utilize α_4 integrin-mediated extravasation. By blocking α_4 integrins, natalizumab attenuates the movement of mononuclear leukocytes to the central nervous system (which utilizes $\alpha_4\beta_1$) and to other inflamed tissues such as the small intestine (which requires $\alpha_4\beta_7$) [97, 98]. While natalizumab has received considerable attention for its clinical efficacy [99, 100], serious concerns were raised when its use was associated with JC virus-induced progressive multifocal leukoencephalopathy (PML) [101]. A follow up study examining 3417 patients who had received natalizumab showed that PML was truly a rare complication with only the original three index cases having confirmed disease; relative risk for developing PML after receiving natalizumab was estimated to be 1 in 1000 patients [102]. Thus, there is renewed hope that this agent will be an effective and relatively safe agent for multiple sclerosis and inflammatory bowel disease.

Selectins

The selectin group of CAMs consists of single-chain transmembrane glycoproteins that share properties with C-type lectins and are, like integrins, involved in the mediation of leukocyte rolling along the endothelium [103]. Selectins are involved in the first step in transmigration of leukocytes from the circulation into the surrounding tissue. Because selectin expression is considered a primary event in the inflammatory response, it represents an attractive target for therapeutic intervention. Unlike integrins and IgSF members, selectin contribution to the immune synapse is not well known nor has it been demonstrated that they significantly contribute to costimulatory signaling. Therefore, targeted selectin treatments likely principally involve leukocyte trafficking. The selectin family of CAMs consists of three structurally related calcium-dependent carbohydrate binding proteins: P-, E- and L- selectin. P-selectin is normally housed in the granules of platelets and endothelial cells and is rapidly inducible in response to inflammatory signals. E-selectin is an intermediate inducible selectin found primarily on activated vascular endothelial cells. L selectin is constitutively expressed on the surface of several leukocyte subtypes including neutrophils, a subset of natural killer cells, monocytes and the majority of circulating B and T cells.

As illustrated in Table 4, pre-clinical models show that blocking selectin activity, not surprisingly, affects the accumulation of leukocytes in various tissues. In earlier studies, mAbs were chiefly employed to block selectin function. Over time, there has been increased understanding that all selectins recognize the carbohydrate structure sLe^x [104], and consequently, the pharmaceutical industry targeted this moiety in the development of newer generation selectin inhibitors. More recently, attention has shifted towards the development of low-molecular-weight selectin inhibitors to facilitate increased synthesis, affinity and oral availability [105]. As with other anti-adhesion therapies, clinical use of selectin inhibitors has met variable success. Although pre-clinical studies in asthma, ischemia-reperfusion injury, psoriasis and myocardial infarction were promising, results of clinical trials with monoclonal antibodies against P-, E- and L- selectins were disappointing. [106]. With the transition to the sLe^x mimetics, clinical results have appeared more encouraging. A new class of compounds called efomycines, which are fermentation by-products of *Streptomyces* BS1261, have shown promising results in experimental models of cutaneous inflammation [107].

Conclusions

The potential of anti-adhesion therapy lies in a more specific targeting of the immune response by interfering with the movement of leukocytes from the circulation into tissues. Although blocking leukocyte trafficking may be the primary effect, it is also likely that interfering with the normal immune synapse is another important reason that anti-adhesion therapy is effective. While there has been considerable interest in IgSF and selectin families of adhesion molecules, the primary focus has centered on the integrin family of adhesion molecules, especially LFA-1 and α_4 integrins, which

Table 4 - Selected pre-clinical and clinical studies targeting selectin members.

Selectin target	Condition treated	Therapy	Treatment outcome	Stage of development	Ref.
P-, E-, L-Selectin	Asthma	Bimosiamose (sLe^x mimetic)	Attenuates late asthma reaction	Phase II	[123]
	Psoriasis	Bimosiamose	Reduces severe psoriatic lesions	Phase II	[106]
	Renal transplantation	Bimosiamose	Prevents allograft rejection	Pre-clinical	[124]
	Psoriasis	*Efomycine* (sLe^x mimetic)	Decreases psoriatic inflammation	Pre-clinical	[125]
	Myocardial infarction	Efomycine	Decreased thrombus formation, myocardial infarction and I-R injury	Pre-clinical	[126]
	I-R following cardiopulmonary bypass	CY1503	Improved cardiac and pulmonary function	Pre-clinical (lamb)	[127]
	I-R following pulmonary thrombo-endarterectomy	CY1503	Decreased I-R injury	Phase II	[128]
	Peritonitis	OC229648 (sLe^x mimetic)	Reduces mononuclear cell infiltration	Pre-clinical	[105]
P- and E-Selectin	Stroke	HuEP5C7 (humanized mAb)	Reduces PMN infiltration and stroke volume	Pre-clinical (non-human primate)	[129]
E-Selectin	Psoriasis	CDP850 (humanized mAb)	No benefit	Phase II	[130]
P- and L-Selectin	Myocardial infarction	rPSGL-Ig	No benefit	Phase II (stopped)	[106]
P-Selectin	I-R injury	CY1747 (mAb)	Reduces PMN infiltration	Pre-clinical	[131]
	Glomerulonephritis	CY1747	Inhibits lipoxin A4 in renal disease but doesn't attenuate PMN infiltration	Pre-clinical	[132]
L-Selectin	Myocardial I-R injury	DREG200	Decreases myocardiac necrosis and endothelial dysfunction	Pre-clinical	[131]

have achieved the most success and attention. Targeting LFA-1 has appeared in preclinical studies to be most effective when used as an adjunctive therapy with other immunomodulatory agents. While effective for the treatment of psoriasis, it has yet to find an effective routine application in transplantation or other immune-mediated disorders. Use of natilizumab, an effective neutralizing agent of α_4 integrins, has been particularly promising for the treatment of multiple sclerosis and inflammatory bowel disease. It must be hoped that with expanding use, growing clinical experience and more selective targeting of inflammatory conditions, anti-adhesion therapies will likely assume useful primary or adjunctive roles in the clinics of the future.

References

1 Barclay A (2003) Membrane proteins with immunoglobulin-like domains – a master superfamily of interaction molecules. *Semin Immunol* 15: 215–223

2 Enlimomab Acute Stroke Trial Investigators (2001) Use of anti-ICAM-1 therapy in ischemic stroke: results of the Enlimomab Acute Stroke Trial. *Neurology* 57: 1428–1434

3 Spitzer TR, McAfee SL, Dey BR, Colby C, Hope J, Grossberg H, Preffer F, Shaffer J, Alexander SI, Sachs DH et al (2003) Nonmyeloablative haploidentical stem-cell transplantation using anti-CD2 monoclonal antibody (MEDI-507)-based conditioning for refractory hematologic malignancies. *Transplantation* 75: 1748–1751

4 Yonekawa K, Harlan JM (2005) Targeting leukocyte integrins in human diseases. *J Leukoc Biol* 77: 129–140

5 Nicolls MR, Gill RG (2006) LFA-1 (CD11a) as a therapeutic target. *Am J Transplant* 6: 27–36

6 Davignon D, Martz E, Reynolds T, Kurzinger K, Springer TA (1981) Lymphocyte function-associated antigen 1 (LFA-1): a surface antigen distinct from Lyt-2,3 that participates in T lymphocyte-mediated killing. *Proc Natl Acad Sci USA* 78: 4535–4539

7 Sanchez-Madrid F, Krensky AM, Ware CF, Robbins E, Strominger JL, Burakoff SJ, Springer TA (1982) Three distinct antigens associated with human T-lymphocyte-mediated cytolysis: LFA-1, LFA-2, and LFA-3. *Proc Natl Acad Sci USA* 79: 7489–7493

8 Ostermann G, Weber KS, Zernecke A, Schroder A, Weber C (2002) JAM-1 is a ligand of the beta(2) integrin LFA-1 involved in transendothelial migration of leukocytes. *Nat Immunol* 3: 151–158

9 Springer TA (1994) Traffic signals for lymphocyte recirculation and leukocyte emigration: The multistep paradigm. *Cell* 76: 301–314

10 Bachmann MF, McKall-Faienza K, Schmits R, Bouchard D, Beach J, Speiser DE, Mak TW, Ohashi PS (1997) Distinct roles for LFA-1 and CD28 during activation of naive T cells: adhesion *versus* costimulation. *Immunity* 7: 549–557

11 Grakoui A, Bromley SK, Sumen C, Davis MM, Shaw AS, Allen PM, Dustin ML (1999)

The immunological synapse: a molecular machine controlling T cell activation. *Science* 285: 221–227
12 Lee KH, Holdorf AD, Dustin ML, Chan AC, Allen PM, Shaw AS (2002) T cell receptor signaling precedes immunological synapse formation. *Science* 295: 1539–1542
13 Monks CR, Freiberg BA, Kupfer H, Sciaky N, Kupfer A (1998) Three-dimensional segregation of supramolecular activation clusters in T cells. *Nature* 395: 82–86
14 Qi SY, Groves JT, Chakraborty AK (2001) Synaptic pattern formation during cellular recognition. *Proc Natl Acad Sci USA* 98: 6548–6553
15 Freiberg BA, Kupfer H, Maslanik W, Delli J, Kappler J, Zaller DM, Kupfer A (2002) Staging and resetting T cell activation in SMACs. *Nat Immunol* 3: 911–917
16 Huang J, Lo PF, Zal T, Gascoigne NR, Smith BA, Levin SD, Grey HM (2002) CD28 plays a critical role in the segregation of PKC theta within the immunologic synapse. *Proc Natl Acad Sci USA* 99: 9369–9373
17 Pribila JT, Quale AC, Mueller KL, Shimizu Y (2004) Integrins and T cell-mediated immunity. *Annu Rev Immunol* 22: 157–180
18 Sanders ME, Makgoba MW, Sharrow SO, Stephany D, Springer TA, Young HA, Shaw S (1988) Human memory T lymphocytes express increased levels of three cell adhesion molecules (LFA-3, CD2, and LFA-1) and three other molecules (UCHL1, CDw29, and Pgp-1) and have enhanced IFN-gamma production. *J Immunol* 140: 1401–1407
19 Lee WT, Vitetta ES (1991) The differential expression of homing and adhesion molecules on virgin and memory T cells in the mouse. *Cell Immunol* 132: 215–222
20 Okumura M, Fujii Y, Takeuchi Y, Inada K, Nakahara K, Matsuda H (1993) Age-related accumulation of LFA-1^{high} cells in a CD8^{+}CD45RA^{high} T cell population. *Eur J Immunol* 23: 1057–1063
21 Moy VT, Brian AA (1992) Signaling by lymphocyte function-associated antigen 1 (LFA-1) in B cells: enhanced antigen presentation after stimulation through LFA-1. *J Exp Med* 175: 1–7
22 Howard DR, Eaves AC, Takei F (1986) Lymphocyte function-associated antigen (LFA-1) is involved in B cell activation. *J Immunol* 136: 4013–4018
23 Fischer A, Durandy A, Sterkers G, Griscelli C (1986) Role of the LFA-1 molecule in cellular interactions required for antibody production in humans. *J Immunol* 136: 3198–3203
24 Owens T (1991) A role for adhesion molecules in contact-dependent T help for B cells. *Eur J Immunol* 21: 979–983
25 Metzler B, Gfeller P, Bigaud M, Li J, Wieczorek G, Heusser C, Lake P, Katopodis A (2004) Combinations of anti-LFA-1, everolimus, anti-CD40 ligand, and allogeneic bone marrow induce central transplantation tolerance through hemopoietic chimerism, including protection from chronic heart allograft rejection. *J Immunol* 173: 7025–7036
26 Yamagami S, Isobe M, Yamagami H, Hori J, Tsuru T (1997) Rejection mechanism and immunosuppression by FK 506 and anti-leukocyte function associated antigen-1 antibody in concordant corneal xenotransplantation. *Transplant Proc* 29: 943–944

27 Rayat GR, Gill RG (2005) Indefinite survival of neonatal porcine islet xenografts by simultaneous targeting of LFA-1 and CD154 or CD45RB. *Diabetes* 54: 443–451
28 Kavanaugh AF, Lightfoot E, Lipsky PE, Oppenheimer-Marks N (1991) Role of CD11/CD18 in adhesion and transendothelial migration of T cells. Analysis utilizing CD18-deficient T cell clones. *J Immunol* 146: 4149–4156
29 Hamann A, Jablonski-Westrich D, Duijvestijn A, Butcher EC, Baisch H, Harder R, Thiele HG (1988) Evidence for an accessory role of LFA-1 in lymphocyte-high endothelium interaction during homing. *J Immunol* 140: 693–699
30 Warnock RA, Askari S, Butcher EC, von Andrian UH (1998) Molecular mechanisms of lymphocyte homing to peripheral lymph nodes. *J Exp Med* 187: 205–216
31 Constantin G, Majeed M, Giagulli C, Piccio L, Kim JY, Butcher EC, Laudanna C (2000) Chemokines trigger immediate beta2 integrin affinity and mobility changes: differential regulation and roles in lymphocyte arrest under flow. *Immunity* 13: 759–769
32 Andrew DP, Spellberg JP, Takimoto H, Schmits R, Mak TW, Zukowski MM (1998) Transendothelial migration and trafficking of leukocytes in LFA-1-deficient mice. *Eur J Immunol* 28: 1959–1969
33 Belperio JA, Keane MP, Burdick MD, Gomperts B, Xue YY, Hong K, Mestas J, Ardehali A, Mehrad B, Saggar R et al (2005) Role of CXCR2/CXCR2 ligands in vascular remodeling during bronchiolitis obliterans syndrome. *J Clin Invest* 115: 1150–1162
34 Berlin PJ, Bacher JD, Sharrow SO, Gonzalez C, Gress RE (1992) Monoclonal antibodies against human T cell adhesion molecules – Modulation of immune function in nonhuman primates. *Transplantation* 53: 840–849
35 Vugmeyster Y, Kikuchi T, Lowes MA, Chamian F, Kagen M, Gilleaudeau P, Lee E, Howell K, Bodary S, Dummer W et al (2004) Efalizumab (anti-CD11a)-induced increase in peripheral blood leukocytes in psoriasis patients is preferentially mediated by altered trafficking of memory $CD8^+$ T cells into lesional skin. *Clin Immunol* 113: 38–46
36 Van Seventer GA, Shimizu Y, Horgan KJ, Shaw S (1990) The LFA-1 ligand ICAM-1 provides an important costimulatory signal for T cell receptor-mediated activation of resting T cells. *J Immunol* 144: 4579–4586
37 Zuckerman LA, Pullen L, Miller J (1998) Functional consequences of costimulation by ICAM-1 on IL-2 gene expression and T cell activation. *J Immunol* 160: 3259–3268
38 Cai Z, Brunmark A, Jackson MR, Loh D, Peterson PA, Sprent J (1996) Transfected Drosophilia cells as a probe for defining the minimal requirements for stimulating unprimed $CD8^+$ T cells. *Proc Natl Acad Sci USA* 93: 14736–14741
39 Abraham C, Griffith J, Miller J (1999) The dependence for leukocyte function-associated antigen-1/ICAM-1 interactions in T cell activation cannot be overcome by expression of high density TCR ligand. *J Immunol* 162: 4399–4405
40 Ragazzo JL, Ozaki ME, Karlsson L, Peterson PA, Webb SR (2001) Costimulation *via* lymphocyte function-associated antigen 1 in the absence of CD28 ligation promotes anergy of naive $CD4^+$ T cells. *Proc Natl Acad Sci USA* 98: 241–246
41 Bleijs DA, van Duijnhoven GC, van Vliet SJ, Thijssen JP, Figdor CG, van Kooyk Y (2001) A single amino acid in the cytoplasmic domain of the beta 2 integrin lymphocyte

function-associated antigen-1 regulates avidity-dependent inside-out signaling. *J Biol Chem* 276: 10338–10346

42 Chatila TA, Geha RS, Arnaout MA (1989) Constitutive and stimulus-induced phosphorylation of CD11/CD18 leukocyte adhesion molecules. *J Cell Biol* 109: 3435–3444

43 Rovere P, Inverardi L, Bender JR, Pardi R (1996) Feedback modulation of ligand-engaged alpha L/beta 2 leukocyte integrin (LFA-1) by cyclic AMP-dependent protein kinase. *J Immunol* 156: 2273–2279

44 Wulfing C, Sjaastad MD, Davis MM (1998) Visualizing the dynamics of T cell activation: intracellular adhesion molecule 1 migrates rapidly to the T cell/B cell interface and acts to sustain calcium levels. *Proc Natl Acad Sci USA* 95: 6302–6307

45 Bianchi E, Denti S, Granata A, Bossi G, Geginat J, Villa A, Rogge L, Pardi R (2000) Integrin LFA-1 interacts with the transcriptional co-activator JAB1 to modulate AP-1 activity. *Nature* 404: 617–621

46 Perez OD, Mitchell D, Jager GC, South S, Murriel C, McBride J, Herzenberg LA, Kinoshita S, Nolan GP (2003) Leukocyte functional antigen 1 lowers T cell activation thresholds and signaling through cytohesin-1 and Jun-activating binding protein 1. *Nat Immunol* 4: 1083–1092

47 Nicolls MR, Coulombe M, Yang H, Bolwerk A, Gill RG (2000) Anti-LFA-1 therapy induces long-term islet allograft acceptance in the absence of IFN-gamma or IL-4. *J Immunol* 164: 3627–3634

48 Nicolls MR, Coulombe M, Beilke J, Gelhaus HC, Gill RG (2002) CD4–dependent generation of dominant transplantation tolerance induced by simultaneous perturbation of CD154 and LFA-1 pathways. *J Immunol* 169: 4831–4839

49 Salomon B, Bluestone JA (1998) LFA-1 interaction with ICAM-1 and ICAM-2 regulates Th2 cytokine production. *J Immunol* 161: 5138–5142

50 Jenks SA, Eisfelder BJ, Miller J (2005) LFA-1 co-stimulation inhibits Th2 differentiation by down-modulating IL-4 responsiveness. *Int Immunol* 17: 315–323

51 Luksch CR, Winqvist O, Ozaki ME, Karlsson L, Jackson MR, Peterson PA, Webb SR (1999) Intercellular adhesion molecule-1 inhibits interleukin 4 production by naive T cells. *Proc Natl Acad Sci USA* 96: 3023–3028

52 Da Silva M, Petruzzo P, Virieux S, Tiollier J, Badet L, Martin X (2001) A primate model of renal ischemia-reperfusion injury for preclinical evaluation of the antileukocyte function associated antigen 1 monoclonal antibody odulimonab. *J Urol* 166: 1915–1919

53 Cavazzana-Calvo M, Sarnacki S, Haddad E, De Coene C, Calise D, Yvon E, Cerf-Bensussan N, Fischer A (1995) Prevention of bone marrow and cardiac graft rejection in an H-2 haplotype disparate mouse combination by an anti-LFA-1 antibody. *Transplantation* 59: 1576–1582

54 Bashuda H, Takazawa K, Tamatani T, Miyasaka M, Yagita H, Okumura K (1996) Induction of persistent allograft tolerance in the rat by combined treatment with anti-leukocyte function-associated antigen-1 and anti-intercellular adhesion molecule-1 monoclonal antibodies, donor-specific transfusion, and FK506. *Transplantation* 62: 117–122

55 Miwa S, Isobe M, Suzuki J, Makuuchi M, Miyasaka M, Yamazaki S, Kawasaki S (1997) Effect of anti-intercellular adhesion molecule-1 and anti-leukocyte function associated antigen-1 monoclonal antibodies on rat-to-mouse cardiac xenograft rejection. *Surgery* 121: 681–689

56 Isobe M, Yagita H, Okumura K, Ihara A (1992) Specific acceptance of cardiac allograft after treatment with antibodies to ICAM-1 and LFA-1. *Science* 255: 1125–1127

57 Horimoto H, Ito T, Hayashi T, Miyasaka M, Nozawa M (1998) Transplantation tolerance by a combined therapy with sulfatide, anti-LFA-1/ICAM-1 monoclonal antibodies and FK506 in rat cardiac transplantation. *Transpl Int* 11 (Suppl 1): S310–312

58 Harrison PC, Madwed JB (1999) Anti-LFA-1 alpha reduces the dose of cyclosporin A needed to produce immunosuppression in heterotopic cardiac transplanted rats. *J Heart Lung Transplant* 18: 279–284

59 Corbascio M, Mahanty H, Osterholm C, Qi Z, Pearson TC, Larsen CP, Freise CE, Ekberg H (2002) Anti-lymphocyte function-associated antigen-1 monoclonal antibody inhibits CD40 ligand-independent immune responses and prevents chronic vasculopathy in CD40 ligand-deficient mice. *Transplantation* 74: 35–41

60 Poston RS, Robbins RC, Chan B, Simms P, Presta L, Jardieu P, Morris RE (2000) Effects of humanized monoclonal antibody to rhesus CD11a in rhesus monkey cardiac allograft recipients. *Transplantation* 69: 2005–2013

61 Corbascio M, Ekstrand H, Osterholm C, Qi Z, Simanaitis M, Larsen CP, Pearson TC, Riesbeck K, Ekberg H (2002) CTLA4Ig combined with anti-LFA-1 prolongs cardiac allograft survival indefinitely. *Transpl Immunol* 10: 55–61

62 Morikawa M, Brazelton TR, Berry GJ, Morris RE (2001) Prolonged inhibition of obliterative airway disease in murine tracheal allografts by brief treatment with anti-leukocyte function-associated antigen-1 (CD11a) monoclonal antibody. *Transplantation* 71: 1616–1621

63 Murakawa T, Kerklo MM, Zamora MR, Wei Y, Gill RG, Henson PM, Grover FL, Nicolls MR (2005) Simultaneous LFA-1 and CD40 ligand antagonism prevents airway remodeling in orthotopic airway transplantation: implications for the role of respiratory epithelium as a modulator of fibrosis. *J Immunol* 174: 3869–3879

64 Nakao Y, Mackinnon SE, Strasberg SR, Hertl MC, Isobe M, Susskind BM, Mohanakumar T, Hunter DA (1995) Immunosuppressive effect of monoclonal antibodies to ICAM-1 and LFA-1 on peripheral nerve allograft in mice. *Microsurgery* 16: 612–620

65 Genden EM, Mackinnon SE, Yu S, Flye MW (1998) Induction of donor-specific tolerance to rat nerve allografts with portal venous donor alloantigen and anti-ICAM-1/LFA-1 monoclonal antibodies. *Surgery* 124: 448–456

66 Larsson LC, Corbascio M, Widner H, Pearson TC, Larsen CP, Ekberg H (2002) Simultaneous inhibition of B7 and LFA-1 signaling prevents rejection of discordant neural xenografts in mice lacking CD40L. *Xenotransplantation* 9: 68–76

67 Isobe M, Suzuki J, Yamazaki S, Sekiguchi M (1996) Acceptance of primary skin graft after treatment with anti-intercellular adhesion molecule-1 and anti-leukocyte function-associated antigen-1 monoclonal antibodies in mice. *Transplantation* 62: 411–413

68 He Y, Mellon J, Apte R, Niederkorn JY (1994) Effect of LFA-1 and ICAM-1 antibody treatment on murine corneal allograft survival. *Invest Ophthalmol Vis Sci* 35: 3218–3225
69 Hori J, Isobe M, Yamagami S, Mizuochi T, Tsuru T (1997) Specific immunosuppression of corneal allograft rejection by combination of anti-VLA-4 and anti-LFA-1 monoclonal antibodies in mice. *Exp Eye Res* 65: 89–98
70 Kato Y, Yamataka A, Yagita H, Bashuda H, Okumura K, Miyano T (1995) Prevention of fetal bowel allograft rejection by combined treatment with anti-ICAM-1 and anti-LFA-1 antibodies. *J Pediatr Surg* 30: 1093–1097
71 Bowles MJ, Pockley AG, Wood RF (2000) Effect of anti-LFA-1 monoclonal antibody on rat small bowel allograft survival and circulating leukocyte populations. *Transpl Immunol* 8: 75–80
72 Gotoh M, Fukuzaki T, Monden M, Dono K, Kanai T, Yagita H, Okumura K, Mori T (1994) A potential immunosuppressive effect of anti-lymphocyte function-associated antigen-1 monoclonal antibody on islet transplantation. *Transplantation* 57: 123–126
73 Nishihara M, Gotoh M, Ohzato H, Ohta Y, Luo Z, Dono K, Umeshita K, Sakon M, Monden M, Yagita H et al (1997) Awareness of donor alloantigens in antiadhesion therapy induces antigen-specific unresponsiveness to islet allografts. *Transplantation* 64: 965–970
74 Arai K, Sunamura M, Wada Y, Takahashi M, Kobari M, Kato K, Yagita H, Okumura K, Matsuno S (1999) Preventing effect of anti-ICAM-1 and anti-LFA-1 monoclonal antibodies on murine islet allograft rejection. *Int J Pancreatol* 26: 23–31
75 Nicolls MR, Coulombe M, Diamond AS, Beilke J, Gill RG (2002) Interferon-gamma is not a universal requirement for islet allograft survival. *Transplantation* 74: 472–477
76 Grochowiecki T, Gotoh M, Dono K, Takeda Y, Sakon M, Yagita H, Okumura K, Miyasaka M, Monden M (2000) Induction of unresponsiveness to islet xenograft by MMC treatment of graft and blockage of LFA-1/ICAM-1 pathway. *Transplantation* 69: 1567–1571
77 Berney T, Pileggi A, Molano RD, Poggioli R, Zahr E, Ricordi C, Inverardi L (2003) The effect of simultaneous CD154 and LFA-1 blockade on the survival of allogeneic islet grafts in nonobese diabetic mice. *Transplantation* 76: 1669–1674
78 Blazar BR, Taylor PA, Panoskaltsis-Mortari A, Gray GS, Vallera DA (1995) Coblockade of the LFA1:ICAM and CD28/CTLA4:B7 pathways is a highly effective means of preventing acute lethal graft-*versus*-host disease induced by fully major histocompatibility complex-disparate donor grafts. *Blood* 85: 2607–2618
79 Wang Y, Gao D, Lunsford KE, Frankel WL, Bumgardner GL (2003) Targeting LFA-1 synergizes with CD40/CD40L blockade for suppression of both CD4-dependent and CD8-dependent rejection. *Am J Transplant* 3: 1251–1258
80 Guerette B, Skuk D, Celestin F, Huard C, Tardif F, Asselin I, Roy B, Goulet M, Roy R, Entman M et al (1997) Prevention by anti-LFA-1 of acute myoblast death following transplantation. *J Immunol* 159: 2522–2531
81 Hasegawa Y, Yokono K, Taki T, Amano K, Tominaga Y, Yoneda R, Yagi N, Maeda S,

Yagita H, Okumura K et al (1994) Prevention of autoimmune insulin-dependent diabetes in non-obese diabetic mice by anti-LFA-1 and anti-ICAM-1 mAb. *Int Immunol* 6: 831–838

82 Chowdhury SA, Nagata M, Yamada K, Nakayama M, Chakrabarty S, Jin Z, Kotani R, Yokono K (2002) Tolerance mechanisms in murine autoimmune diabetes induced by anti-ICAM-1/LFA-1 mAb and anti-CD8 mAb. *Kobe J Med Sci* 48: 167–175

83 Cavazzana-Calvo M, Bordigoni P, Michel G, Esperou H, Souillet G, Leblanc T, Stephan JL, Vannier JP, Mechinaud F, Reiffers J et al (1996) A phase II trial of partially incompatible bone marrow transplantation for high-risk acute lymphoblastic leukaemia in children: prevention of graft rejection with anti-LFA-1 and anti-CD2 antibodies. Societe Francaise de Greffe de Moelle Osseuse. *Br J Haematol* 93: 131–138

84 Maraninchi D, Mawas C, Stoppa AM, Gaspard MH, Marit G, Van Ekthoven A, Reiffers J, Olive D, Hirn M, Delaage M et al (1989) Anti LFA1 monoclonal antibody for the prevention of graft rejection after T cell-depleted HLA-matched bone marrow transplantation for leukemia in adults. *Bone Marrow Transplant* 4: 147–150

85 Le Mauff B, Hourmant M, Rougier JP, Hirn M, Dantal J, Baatard R, Cantarovich D, Jacques Y, Soulillou JP (1991) Effect of anti-LFA1 (CD11a) monoclonal antibodies in acute rejection in human kidney transplantation. *Transplantation* 52: 291–296

86 Hourmant M, Bedrossian J, Durand D, Lebranchu Y, Renoult E, Caudrelier P, Buffet R, Soulillou JP (1996) A randomized multicenter trial comparing leukocyte function-associated antigen-1 monoclonal antibody with rabbit antithymocyte globulin as induction treatment in first kidney transplantations. *Transplantation* 62: 1565–1570

87 Benfield MR, Tejani A, Harmon WE, McDonald R, Stablein DM, McIntosh M, Rose S (2005) A randomized multicenter trial of OKT3 mAbs induction compared with intravenous cyclosporine in pediatric renal transplantation. *Pediatr Transplant* 9: 282–292

88 Werther WA, Gonzalez TN, O'Connor SJ, McCabe S, Chan B, Hotaling T, Champe M, Fox JA, Jardieu PM, Berman PW et al (1996) Humanization of an anti-lymphocyte function-associated antigen (LFA)-1 monoclonal antibody and reengineering of the humanized antibody for binding to rhesus LFA-1. *J Immunol* 157: 4986–4995

89 Vincenti F, Mendez R, Pescovitz M, Rajagopalan PR, Wilkinson AH, Butt K, Laskow D, Slakey DP, Lorber MI, Garg JP et al (2007) A Phase I/II randomized open-label multicenter trial of efalizumab, a humanized anti-CD11a, anti-LFA-1 in renal transplantation. *Am J Transplant* 7: 1770–1777

90 Gabrijelcic J, Acuna A, Profita M, Paterno A, Chung KF, Vignola AM, Rodriguez-Roisin R (2003) Neutrophil airway influx by platelet-activating factor in asthma: role of adhesion molecules and LTB4 expression. *Eur Respir J* 22: 290–297

91 Xu B, Wagner N, Pham LN, Magno V, Shan Z, Butcher EC, Michie SA (2003) Lymphocyte homing to bronchus-associated lymphoid tissue (BALT) is mediated by L-selectin/PNAd, alpha4beta1 integrin/VCAM-1, and LFA-1 adhesion pathways. *J Exp Med* 197: 1255–1267

92 Gauvreau GM, Becker AB, Boulet LP, Chakir J, Fick RB, Greene WL, Killian KJ, O'Byrne P M, Reid JK, Cockcroft DW (2003) The effects of an anti-CD11a mAb, efali-

zumab, on allergen-induced airway responses and airway inflammation in subjects with atopic asthma. J Allergy *Clin Immunol* 112: 331–338
93 Chou YK, Edwards DM, Weinberg AD, Vandenbark AA, Kotzin BL, Fontenot AP, Burrows GG (2005) Activation pathways implicate anti-HLA-DP and anti-LFA-1 antibodies as lead candidates for intervention in chronic berylliosis. *J Immunol* 174: 4316–4324
94 Leonardi CL (2003) Efalizumab: an overview. *J Am Acad Dermatol* 49: S98–104
95 Stern RS (2003) A promising step forward in psoriasis therapy. *JAMA* 290: 3133–3135
96 Mittelbrunn M, Molina A, Escribese MM, Yanez-Mo M, Escudero E, Ursa A, Tejedor R, Mampaso F, Sanchez-Madrid F (2004) VLA-4 integrin concentrates at the peripheral supramolecular activation complex of the immune synapse and drives T helper 1 responses. *Proc Natl Acad Sci USA* 101: 11058–11063
97 von Andrian UH, Engelhardt B (2003) Alpha4 integrins as therapeutic targets in autoimmune disease. *N Engl J Med* 348: 68–72
98 Engelhardt B, Ransohoff RM (2005) The ins and outs of T-lymphocyte trafficking to the CNS: anatomical sites and molecular mechanisms. *Trends Immunol* 26: 485–495
99 Polman CH, O'Connor PW, Havrdova E, Hutchinson M, Kappos L, Miller DH, Phillips JT, Lublin FD, Giovannoni G, Wajgt A et al (2006) A randomized, placebo-controlled trial of natalizumab for relapsing multiple sclerosis. *N Engl J Med* 354: 899–910
100 Rudick RA, Stuart WH, Calabresi PA, Confavreux C, Galetta SL, Radue EW, Lublin FD, Weinstock-Guttman B, Wynn DR, Lynn F et al (2006) Natalizumab plus interferon beta-1a for relapsing multiple sclerosis. *N Engl J Med* 354: 911–923
101 Van Assche G, Van Ranst M, Sciot R, Dubois B, Vermeire S, Noman M, Verbeeck J, Geboes K, Robberecht W, Rutgeerts P (2005) Progressive multifocal leukoencephalopathy after natalizumab therapy for Crohn's disease. *N Engl J Med* 353: 362–368
102 Yousry TA, Major EO, Ryschkewitsch C, Fahle G, Fischer S, Hou J, Curfman B, Miszkiel K, Mueller-Lenke N, Sanchez E et al (2006) Evaluation of patients treated with natalizumab for progressive multifocal leukoencephalopathy. *N Engl J Med* 354: 924–933
103 Ley K (2001) Functions of selectins. *Results Probl Cell Differ* 33: 177–200
104 Foxall C, Watson SR, Dowbenko D, Fennie C, Lasky LA, Kiso M, Hasegawa A, Asa D, Brandley BK (1992) The three members of the selectin receptor family recognize a common carbohydrate epitope, the sialyl Lewis(x) oligosaccharide. *J Cell Biol* 117: 895–902
105 Slee DH, Romano SJ, Yu J, Nguyen TN, John JK, Raheja NK, Axe FU, Jones TK, Ripka WC (2001) Development of potent non-carbohydrate imidazole-based small molecule selectin inhibitors with antiinflammatory activity. *J Med Chem* 44: 2094–2107
106 Ulbrich H, Eriksson EE, Lindbom L (2003) Leukocyte and endothelial cell adhesion molecules as targets for therapeutic interventions in inflammatory disease. *Trends Pharmacol Sci* 24: 640–647
107 Wienrich BG, Krahn T, Schon M, Rodriguez ML, Kramer B, Busemann M, Boehncke

WH, Schon MP (2006) Structure-function relation of efomycines, a family of small-molecule inhibitors of selectin functions. *J Invest Dermatol* 126: 882–889
108 Zhu GD, Arendsen DL, Gunawardana IW, Boyd SA, Stewart AO, Fry DG, Cool BL, Kifle L, Schaefer V, Meuth J et al (2001) Selective inhibition of ICAM-1 and E-selectin expression in human endothelial cells. 2. Aryl modifications of 4–(aryloxy)thieno[2,3–c]pyridines with fine-tuning at C-2 carbamides. *J Med Chem* 44: 3469–3487
109 Salmela K, Wramner L, Ekberg H, Hauser I, Bentdal O, Lins LE, Isoniemi H, Backman L, Persson N, Neumayer HH et al (1999) A randomized multicenter trial of the anti-ICAM-1 monoclonal antibody (enlimomab) for the prevention of acute rejection and delayed onset of graft function in cadaveric renal transplantation: a report of the European Anti-ICAM-1 Renal Transplant Study Group. *Transplantation* 67: 729–736
110 Justicia C, Martin A, Rojas S, Gironella M, Cervera A, Panes J, Chamorro A, Planas AM (2006) Anti-VCAM-1 antibodies did not protect against ischemic damage either in rats or in mice. *J Cereb Blood Flow Metab* 26: 421–432
111 Gumina RJ, el Schultz J, Yao Z, Kenny D, Warltier DC, Newman PJ, Gross GJ (1996) Antibody to platelet/endothelial cell adhesion molecule-1 reduces myocardial infarct size in a rat model of ischemia-reperfusion injury. *Circulation* 94: 3327–3333
112 Murohara T, Delyani JA, Albelda SM, Lefer AM (1996) Blockade of platelet endothelial cell adhesion molecule-1 protects against myocardial ischemia and reperfusion injury in cats. *J Immunol* 156: 3550–3557
113 Isobe M, Suzuki J, Yamazaki S, Yazaki Y, Horie S, Okubo Y, Maemura K, Yazaki Y, Sekiguchi M (1997) Regulation by differential development of Th1 and Th2 cells in peripheral tolerance to cardiac allograft induced by blocking ICAM-1/LFA-1 adhesion. *Circulation* 96: 2247–2253
114 Bloemen PG, Buckley TL, van den Tweel MC, Henricks PA, Redegeld FA, Koster AS, Nijkamp FP (1996) LFA-1, and not Mac-1, is crucial for the development of hyperreactivity in a murine model of nonallergic asthma. *Am J Respir Crit Care Med* 153: 521–529
115 Theien BE, Vanderlugt CL, Eagar TN, Nickerson-Nutter C, Nazareno R, Kuchroo VK, Miller SD (2001) Discordant effects of anti-VLA-4 treatment before and after onset of relapsing experimental autoimmune encephalomyelitis. *J Clin Invest* 107: 995–1006
116 Theien BE, Vanderlugt CL, Nickerson-Nutter C, Cornebise M, Scott DM, Perper SJ, Whalley ET, Miller SD (2003) Differential effects of treatment with a small-molecule VLA-4 antagonist before and after onset of relapsing EAE. *Blood* 102: 4464–4471
117 Dubree NJ, Artis DR, Castanedo G, Marsters J, Sutherlin D, Caris L, Clark K, Keating SM, Beresini MH, Chiu H et al (2002) Selective alpha4beta7 integrin antagonists and their potential as antiinflammatory agents. *J Med Chem* 45: 3451–3457
118 Faxon DP, Gibbons RJ, Chronos NA, Gurbel PA, Sheehan F (2002) The effect of blockade of the CD11/CD18 integrin receptor on infarct size in patients with acute myocardial infarction treated with direct angioplasty: the results of the HALT-MI study. *J Am Coll Cardiol* 40: 1199–1204
119 Targan SR, Feagan BG, Fedorak RN, Lashner BA, Panaccione R, Present DH, Spehl-

mann ME, Rutgeerts PJ, Tulassay Z, Volfova M et al (2007) Natalizumab for the treatment of active Crohn's disease: results of the ENCORE Trial. *Gastroenterology* 132: 1672–1683
120 Ghosh S, Goldin E, Gordon FH, Malchow HA, Rask-Madsen J, Rutgeerts P, Vyhnalek P, Zadorova Z, Palmer T, Donoghue S (2003) Natalizumab for active Crohn's disease. *N Engl J Med* 348: 24–32
121 Gordon FH, Lai CW, Hamilton MI, Allison MC, Srivastava ED, Fouweather MG, Donoghue S, Greenlees C, Subhani J, Amlot PL et al (2001) A randomized placebo-controlled trial of a humanized monoclonal antibody to alpha4 integrin in active Crohn's disease. *Gastroenterology* 121: 268–274
122 Ravensberg AJ, Luijk B, Westers P, Hiemstra PS, Sterk PJ, Lammers JW, Rabe KF (2006) The effect of a single inhaled dose of a VLA-4 antagonist on allergen-induced airway responses and airway inflammation in patients with asthma. *Allergy* 61: 1097–1103
123 Beeh KM, Beier J, Meyer M, Buhl R, Zahlten R, Wolff G (2006) Bimosiamose, an inhaled small-molecule pan-selectin antagonist, attenuates late asthmatic reactions following allergen challenge in mild asthmatics: a randomized, double-blind, placebo-controlled clinical cross-over-trial. *Pulm Pharmacol Ther* 19: 233–241
124 Langer R, Wang M, Stepkowski SM, Hancock WW, Han R, Li P, Feng L, Kirken RA, Berens KL, Dupre B et al (2004) Selectin inhibitor bimosiamose prolongs survival of kidney allografts by reduction in intragraft production of cytokines and chemokines. *J Am Soc Nephrol* 15: 2893–2901
125 Oostingh GJ, Ludwig RJ, Enders S, Gruner S, Harms G, Boehncke WH, Nieswandt B, Tauber R, Schon MP (2007) Diminished lymphocyte adhesion and alleviation of allergic responses by small-molecule- or antibody-mediated inhibition of L-selectin functions. *J Invest Dermatol* 127: 90–97
126 Oostingh GJ, Pozgajova M, Ludwig RJ, Krahn T, Boehncke WH, Nieswandt B, Schon MP (2007) Diminished thrombus formation and alleviation of myocardial infarction and reperfusion injury through antibody- or small-molecule-mediated inhibition of selectin-dependent platelet functions. *Haematologica* 92: 502–512
127 Schermerhorn ML, Tofukuji M, Khoury PR, Phillips L, Hickey PR, Sellke FW, Mayer JE Jr, Nelson DP (2000) Sialyl lewis oligosaccharide preserves cardiopulmonary and endothelial function after hypothermic circulatory arrest in lambs. *J Thorac Cardiovasc Surg* 120: 230–237
128 Kerr KM, Auger WR, Marsh JJ, Comito RM, Fedullo RL, Smits GJ, Kapelanski DP, Fedullo PF, Channick RN, Jamieson SW et al (2000) The use of cylexin (CY-1503) in prevention of reperfusion lung injury in patients undergoing pulmonary thromboendarterectomy. *Am J Respir Crit Care Med* 162: 14–20
129 Mocco J, Choudhri T, Huang J, Harfeldt E, Efros L, Klingbeil C, Vexler V, Hall W, Zhang Y, Mack W et al (2002) HuEP5C7 as a humanized monoclonal anti-E/P-selectin neurovascular protective strategy in a blinded placebo-controlled trial of nonhuman primate stroke. *Circ Res* 91: 907–914
130 Bhushan M, Bleiker TO, Ballsdon AE, Allen MH, Sopwith M, Robinson MK, Clarke

C, Weller RP, Graham-Brown RA, Keefe M et al (2002) Anti-E-selectin is ineffective in the treatment of psoriasis: a randomized trial. *Br J Dermatol* 146: 824–831

131 Lefer DJ (2000) Pharmacology of selectin inhibitors in ischemia/reperfusion states. *Annu Rev Pharmacol Toxicol* 40: 283–294

132 Papayianni A, Serhan CN, Phillips ML, Rennke HG, Brady HR (1995) Transcellular biosynthesis of lipoxin A4 during adhesion of platelets and neutrophils in experimental immune complex glomerulonephritis. *Kidney Int* 47: 1295–1302

E3 ubiquitin ligases and immune tolerance: Targeting the immune synapse from within?

Irene Puga and Fernando Macian

Albert Einstein College of Medicine, Department of Pathology, 1300 Morris Park Avenue, Bronx, NY 10461, USA

T cell anergy

General concept

The success of adaptive immunity relies on the ability to eliminate invading pathogens without eliciting responses against the host. Unique antigen receptors are randomly generated and recognize both self and non-self antigens. Therefore, mechanisms of tolerance must be in place to control the activity of self-reactive lymphocytes. Negative selection in the thymus eliminates most of the developing thymocytes that can recognize self antigens [1], whereas mechanisms of peripheral tolerance prevent the surviving self-reactive cells from engaging in responses against self tissues. Self-reactive T cells can be suppressed by regulatory T cells, and also eliminated by clonal deletion or inactivated by a mechanism known as anergy [2]. In anergic T cells, T cell receptor (TCR) signaling is blocked, and cells become unresponsive to subsequent stimulation events [3–5].

Productive activation of T cells requires the integration of two different signals: the engagement by the TCR of an antigenic peptide presented by the major histocompatibility complex (MHC) at the plasma membrane of antigen-presenting cells (APCs), and the interaction of costimulatory signals. Professional APCs express costimulatory molecules such as B7-1/CD80 and B7-2/CD86, which are ligands for the costimulatory receptor CD28, expressed in T cells [6]. Engagement of the TCR and CD28 leads to T cell activation, which results in increased cytokine production and clonal proliferation. The intensity/strength and duration of the TCR engagement is crucial in determining the nature of T cell responses, and CD28 costimulation can modulate the threshold that allows appropriate signaling for T cell activation [7]. The formation of a mature immune synapse in the contact interface between the T cell and the APC is critical for this process, defining a platform of signaling molecules that reorganize to ensure the efficient activation of T cells [8–10] (Fig. 1).

In the absence of "danger" signals provided by pathogens, APCs will present non-pathogenic antigens with little or no costimulation engagement, inducing a long-lasting state of functional unresponsiveness in T cells, namely anergy [3]

The Immune Synapse as a Novel Target for Therapy, edited by Luis Graca

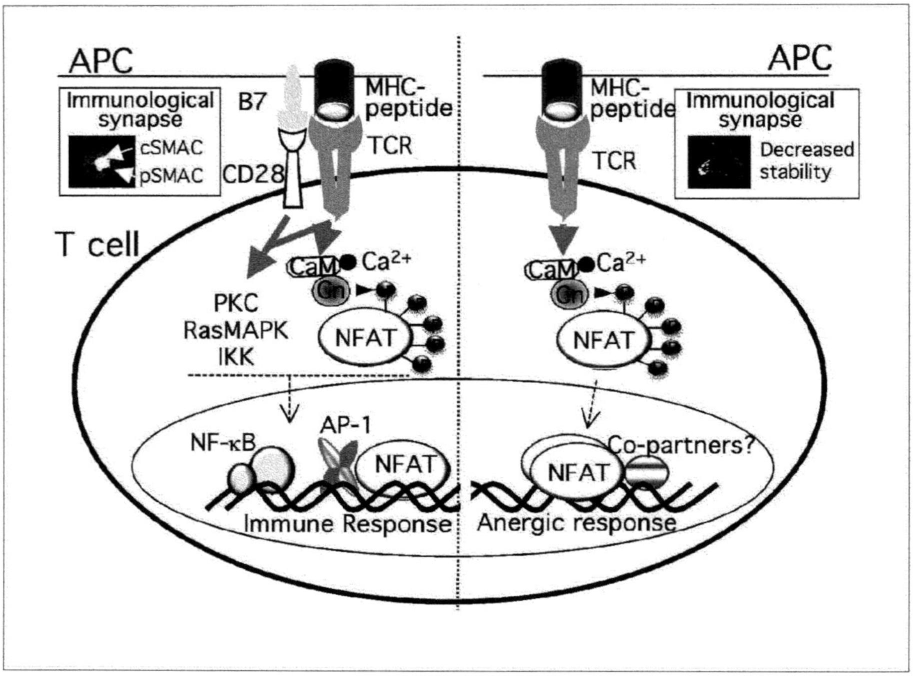

Figure 1

Signaling pathways in activated and anergic T cells.

T cell activation requires engagement of the TCR and CD28. The TCR recognizes a specific peptide presented on the MHC by APCs. This engagement increases the intracellular calcium levels, which leads to the dephosphorylation and activation of the transcription factor NFAT by the calcium/calmodulin (CaM)-dependent phosphatase calcineurin (Cn). B7 proteins (B7.1 and B7.2) bind CD28 and modulate together with signals initiated from TCR engagement different signaling pathways that eventually activate the transcription factors AP-1 and NF-κB. The immune response, characterized by increased cytokine production and cell proliferation, requires the formation of a stable immune synapse in the interface between the APC and the T cell. The immune synapse results from an organized recruitment of signaling molecules and receptors organized in a supramolecular activation cluster (SMAC) arrangement, formed by a central SMAC (c-SMAC), containing the TCR-MHC: peptide complexes and other signaling molecules, surrounded by the peripheral SMAC (p-SMAC), that includes among others the integrin lymphocyte function-associated antigen-1 (LFA-1) interacting with its major co-receptor ICAM-1. In the absence of costimulation, NFAT is preferentially activated leading to the up-regulation of anergy-associated genes, which make anergic cells unresponsive and unable to form stable synapses in subsequent stimulations.

(Fig. 1). The first studies that defined anergy induction in T cells were performed by analyzing the effects of TCR ligation in the absence of full costimulation on T cell clones, either on chemically fixed splenocytes [11], planar lipid membranes containing only MHC class II molecules [12], or on plastic surface coated with anti-CD3 antibodies [13]. While the presence of costimulation can prevent anergy induction in responding T cells [14], the negative ligand for the B7 molecules, the cytotoxic T lymphocyte antigen-4 (CTLA-4), which mediates inhibitory effects opposing the costimulatory CD28 signaling pathway [15–17], may be required for the maintenance of anergy *in vivo* [18, 19].

Induction of anergy

Despite the possibility that different forms of anergy may exist, extensive studies have characterized the process of tolerance in $CD4^+$ T cells in two sequential stages: the induction and the maintenance of anergy [4]. Tolerogenic stimuli cause an increase in intracellular free calcium, which leads to the activation of the nuclear factor of activated T cells (NFAT) transcription factors. Three of the five members of the NFAT family of transcription factors members, NFAT1, 2 and 4, are expressed in T cells as highly phosphorylated cytoplasmic proteins. Engagement of calcium-coupled receptors, such as the TCR, induces an increase in intracellular Ca^{2+} that leads to the dephosphorylation of NFAT proteins by the Ca^{2+}/calmodulin-dependent phosphatase calcineurin and their translocation into the nucleus. Once in the nucleus, NFAT proteins bind specific DNA regulatory regions to control the transcription of different genes [20–22]. During T cell activation, NFAT and activator protein (AP)-1 transcription factors cooperatively regulate the expression of T cell activation-associated genes [20, 23, 24]. However, in response to an anergic stimulus, low and sustained levels of Ca^{2+} mobilization activate NFAT in the absence of its main transcriptional partner, AP-1, driving the induction of a specific program of genes [25] (Fig. 1). The expression of these anergy-associated genes is responsible for the induction of an unresponsive state and for the inhibition of cytokine expression in anergic T cells [26, 27].

Maintenance of anergy

Previous studies had suggested that anergy required the synthesis of new proteins that included dominant-acting repressor molecules [12, 28]. It has been only recently that many of these proteins have been identified and the mechanisms responsible of imposing the anergic state characterized. Some of the non-exclusive mechanisms required to induce a state of functional unresponsiveness are summarized in this section.

TCR engagement-induced signaling is profoundly dampened in anergic T cells. The inactivation or degradation of several signaling molecules downstream of the TCR underlies this effect. Among the anergy-inducing genes, several ubiquitin ligases, including Itch, the gene related to anergy in lymphocytes (GRAIL) and the Casitas B-lineage Lymphoma (Cbl)-b, have been shown to target and ubiquitinate specific proteins, such as the phospholipase C (PLC)-γ1, the protein kinase C (PKC)-θ and the Ras GTPase-activating protein RasGAP, leading to defective signaling and subsequent alterations of the stability of the immune synapse [29].

Recent reports have also shown that anergic T cells up-regulate the expression of diacylglycerol kinase α (DAGKα) [25, 30, 31]. Excessive inactivation of DAG by DAGKα prevents the recruitment of the guanine nucleotide exchange factor RasGRP1, which results in uncoupling of Ras activation from TCR engagement, and thus, in defective activation of mitogen-activated protein kinases (MAPK) in anergic T cells [30, 31].

Reduced cellular proliferation is another hallmark of T cell anergy. An inability to down-regulate the cell cycle inhibitor $p27^{kip1}$ has been described in *in vitro* and *in vivo* tolerized T cells, which leads to defective phosphorylation-induced inactivation of Smad3 and prevents cell cycle progression from G1 to S [32, 33].

Besides impaired TCR signaling, anergic T cells also activate direct mechanisms of repression of cytokine expression. Inhibitory complexes, such as CREB/CREM, have the ability to bind the IL-2 promoter and block transcription [34]. Moreover, the transcriptional repressor Ikaros is expressed in anergic T cells and binds the IL-2 promoter, recruiting histone deacetylases (HDAC) that decreased the acetylation status of histones at the IL-2 promoter. These epigenetic changes cause a stable inhibition of IL-2 expression [35].

This chapter focuses on the role of one of these groups of proteins, the E3 ubiquitin ligases Itch, GRAIL and Cbl-b, in the maintenance of the TCR signaling blockade in anergic T cells.

The immune synapse in anergic T cells

The immune synapse forms at the interface between the T cell and the APC. The duration, dynamics, and strength of signaling events in this crucial structure determine the outcome of the T cell-APC interaction. The synapse is organized into a central cluster, where the TCRs engaging MHC:peptide complexes localize (central supramolecular activation cluster or cSMAC), and a peripheral region, where the LFA-1/ICAM-1 contacts occur (peripheral supramolecular activation cluster or pSMAC) [9]. The stability of the immune synapse is altered in anergic T cells, as it disintegrates with much faster kinetics that in non-anergic T cells (Fig. 1). This defect has been attributed to decreased levels of PLCγ1, which would be a consequence of the activation of E3 ubiquitin ligases [29]. Other studies have also described impaired recruitment of the adaptor protein linker for activation of T cells (LAT) to the immune synapse in anergic T cells, which results from inefficient LAT palmitoylation [36]. Defects in TCR signal transduction in anergic T cells could then respond to alterations in the formation of the immune synapse caused by the actions of E3 ligases, to the defective recruitment of essential components of the T cell signaling complex or even to alterations in the activation-induced restructuring of the cytoskeleton.

The ubiquitin system

Ubiquitination

Ubiquitin is a highly conserved protein of 76 amino acids that can be conjugated to other proteins by a covalent attachment typically to lysine residues (Lys) on the target proteins. Ubiquitin itself can accept this conjugation, which allows the formation of polyubiquitin chains [37]. This post-translational modification has been shown to modulate many different properties of proteins, including stability, localization, activation, conformation and the ability to interact with other proteins [38]. The process of ubiquitination involves three different types of enzymes: E1 or ubiquitin-activating enzyme, E2 or ubiquitin-conjugating enzyme, and E3 or ubiquitin-protein ligase. First, the E1 enzyme forms a high-energy thioester bond between a glycine residue of ubiquitin and the cysteine residue located in the active site of the E1 enzyme in an ATP-dependent process. The activated ubiquitin is then transferred to the cysteine residue of the active site in the E2 enzyme. Finally, the E3 enzyme specifically binds the substrate and transfers the ubiquitin from the E2 enzyme (directly or indirectly) to a Lys residue in the substrate protein [38, 39] (Fig. 2). There are two families of E3 ubiquitin ligases: the RING (really interesting new gene) and the HECT (homologous to E6-AP C-terminal) domain-containing families. Both HECT (Itch and Nedd4) and RING (Cbl-b and GRAIL) ubiquitin E3 ligases are involved in T cell anergy [40]. Ubiquitination is also regulated by ubiquitin receptor proteins, which contain ubiquitin-binding domains that recognize the ubiquitinated proteins and control downstream biochemical processes, and by deubiquitinating enzymes, which remove ubiquitin molecules attached to a target protein [41, 42].

Cellular functions of ubiquitin

The nature of the ubiquitin modification is critical in targeting a specific substrate to a particular cellular function. Target proteins can accept one ubiquitin molecule (in a single Lys residue, i.e., monoubiquitination, or in multiple residues, i.e., oligoubiquitination). Moreover, ubiquitin contains seven Lys residues through which different

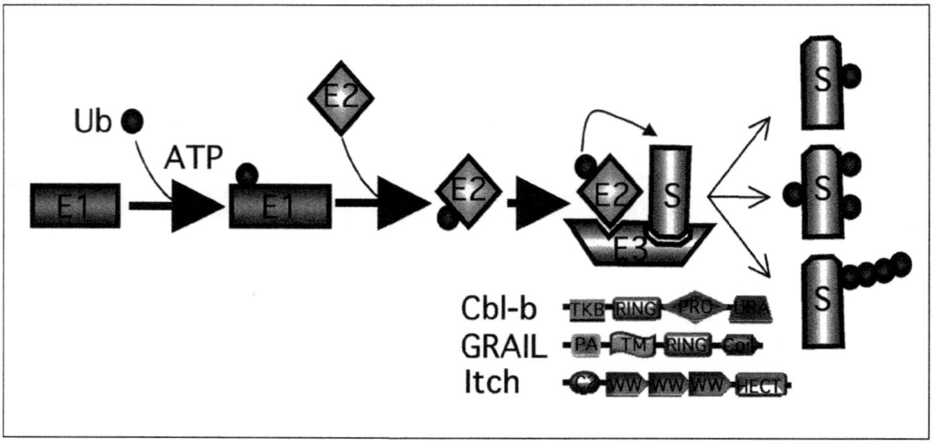

Figure 2
The process of ubiquitination and E3 ligases.
The ubiquitin-activating enzyme (E1) activates free ubiquitin (Ub) in an ATP-dependent process to form a thioester bond. The activated ubiquitin is transferred to the ubiquitin-conjugating enzyme (E2). Commonly, the E2 enzyme interacts specifically with the ubiquitin-ligase enzyme (E3) to transfer the ubiquitin to the protein substrate. The process can lead to mono-, oligo- and polyubiquitination of the substrate. Diagram shows the structural domains of the E3 ubiquitin ligases involved in T cell tolerance. In Cbl-b: tyrosine-kinase binding domain (TKB), RING finger domain, proline-rich sequence (PRO) and ubiquitin binding domain (LZ/UBA). In GRAIL: protease-associated domain (PA), transmembrane segment (TM), coil-coiled domain (Coil) and RING finger domain. In Itch: conserved region related to protein kinase C (C2) domain, dual tryptophan regions (WW domains) and homologous to E6-AP C-terminal (HECT) domain.

types of polyubiquitin chains can be formed. The best-characterized polychains *in vivo* are those that occur through linkage to residues Lys48 and Lys63 [37, 43]. The first identified function of ubiquitin was to target proteins for proteasomal degradation. The ground-breaking studies on the ubiquitin-proteasome system were key in the characterization of an essential pathway of cytosolic protein turnover [44–46]. The targeting of proteins for degradation by the proteasome is generally directed by polyubiquitination with linkage through Lys48 and Lys29. Tagged proteins are recognized by a large multiprotein complex, the 26S proteasome, where the ubiquitin tag is removed and the proteins are subsequently degraded by different protease activities [39].

Besides proteasomal targeting, recent studies have indicated that ubiquitination can regulate many other cellular processes. Examples of these non-proteasomal functions include the role of ubiquitination in modulating tolerance to DNA damage, in the regulation of transcription, in the down-regulation of receptor signaling by targeting to the endocytic pathway, or in the sorting of proteins to specific subcellular compartments [47].

In the immune system ubiquitin is involved in the regulation of the innate and the adaptive responses. Protein ubiquitination has a fundamental role not only in immune tolerance but also in T cell differentiation and the modulation of TCR signaling. For example, ubiquitin regulates the activation of NF-κB, a key transcription factor for the immune response, by tagging the inhibitor molecule bound to NF-κB, IκB. Polyubiquitination of IκB leads to its degradation by the proteasome, releasing NF-κB, which can then be translocated into the nucleus [48]. Ubiquitin also plays a crucial role in TCR downmodulation after antigen engagement. Ubiquitination of the TCR ζ chains leads to sorting of these proteins to the endosomal pathway and their eventual degradation by the lysosomes [49].

E3 ubiquitin ligases and immune tolerance

In T cells, induction of tolerance is characterized not only by alterations in the levels of protein phosphorylation but also by a general increase in total protein ubiquitination [29]. The expression of at least three E3 ubiquitin ligases is up-regulated in anergic T cells: GRAIL, Itch and Cbl-b. Recent studies have unveiled the role of these E3 enzymes in the anergy-associated downmodulation of TCR signaling [50–52].

Cbl-b

The Cbl family of proteins comprises a family of genes related to the viral oncogene v-Cbl, which promotes development of B cell lymphomas. In mammals, three members of this family have been described: c-Cbl, Cbl-b and Cbl-3. Cbl-b is a RING finger-containing E3 ubiquitin ligases expressed in hematopoietic cells, which is involved in the negative regulation of the TCR and other receptors signaling pathway (Fig. 3).

Structure and protein interactions

The structure of Cbl-b is characterized by the presence of a tyrosine kinase binding domain (TKB), a RING domain that confers the E3 ligase catalytic activity, a proline rich domain necessary for protein-protein interaction, and a ubiquitin-associated (UAB)/leucine zipper domain involved in dimerization (Fig. 3A). Cbl-b is recruited to the immune synapse in antigen-stimulated T cells, where it mediates ubiquitina-

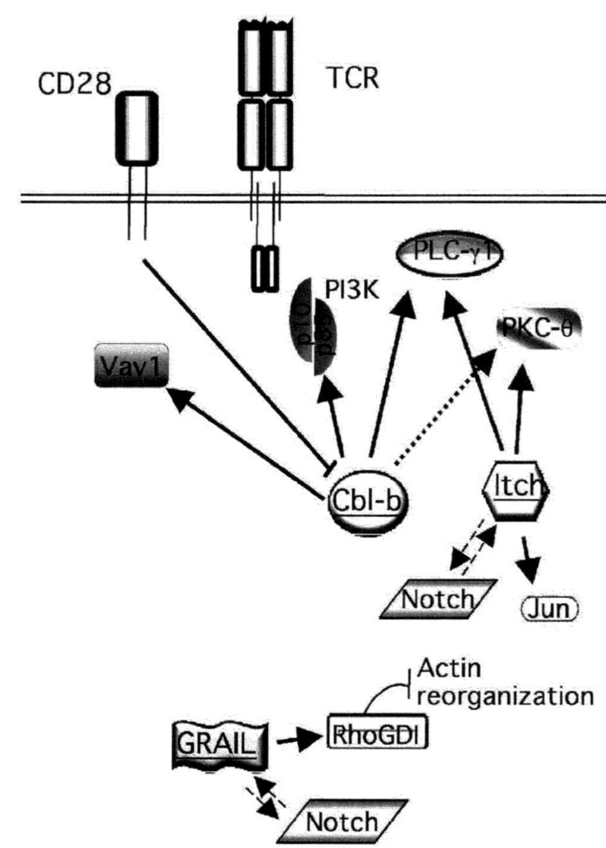

Figure 3

Summary of the E3 ubiquitin ligases substrates involved in T cell tolerance.

The E3 ubiquitin ligases Cbl-b, Itch and GRAIL target different signaling molecules required for T cell activation. Cbl-b binds the p85 subunit of phosphatidylinositol 3-kinase (PI3K), preventing its recruitment to the TCR or CD28. Cbl-b also modulates the function of Vav1, PLC-γ1 and PKC-θ. Itch is responsible for the ubiquitination and subsequent degradation of PLC-γ1 and PKC-θ. Itch also targets the Jun family of transcription factors for degradation and may cooperate with Notch. GRAIL targets the Rho guanine dissociation inhibitor (RhoGDI) stabilizing it by ubiquitination, interfering with TCR-induced actin cytoskeleton reorganization. Notch signaling may also control GRAIL expression.

tion of different substrates [51, 53]. Cbl-b binds and promotes ubiquitination of the p85 regulatory domain of the phosphatidylinositol 3-kinase (PI3K). This modification affects p85 subcellular localization, reducing its recruitment to the immune synapse and, therefore, preventing the interaction of PI3K with CD28 and the TCR ζ chain [54]. Cbl-b also regulates Vav1 activity not by directing its degradation, but rather by regulating the phosphorylation of this adaptor molecule and its recruitment to the immune synapse. By down-regulating Vav1 activity, Cbl-b indirectly controls the activation of Rho family GTPases and the reorganization of the actin cytoskeleton that follows T cell activation [50, 55].

Regulation

Both mRNA and protein levels of Cbl-b are up-regulated during anergy induction [29, 52]. The activation of NFAT proteins promotes the expression of the members of the early growth response protein (Egr) family of transcription factors Egr-2 and Egr-3. These zinc finger transcription factors induce the expression of Cbl-b [56]. At the protein level, Cbl-b is itself a target of ubiquitination and proteasomal degradation, which is promoted by signaling through the costimulatory molecule CD28 [57]. Opposing CD28 effects, CTLA-4 engagement promotes Cbl-b re-expression [58]. Furthermore, TCR engagement also induces phosphorylation of Cbl-b at several tyrosine residues, which regulates Cbl-b interactions with other molecules [59].

Cbl-b deficiency

Cbl-b-deficient mice show hyperactive T cells and develop spontaneous autoimmunity and inflammatory tissue damage [50, 60]. A dramatic increase in proliferative responses and cytokine production by $CD4^+$ T cells in response to antigen, and an enhanced cytolytic activity by $CD8^+$ T cells can also be found in these mice, as well as high titers of anti-DNA antibodies and hyperreactive B cells. Cbl-$b^{-/-}$ T cells do not require CD28 costimulation to effectively proliferate or produce cytokines, whereas the lack of Cbl-b can rescue hyporesponsiveness in T cells that do not express CD28. Cbl-b deficiency causes hyperactivation of Vav1 with increased TCR clustering [50, 51, 60]. Mice with T cells that are deficient in both c-Cbl and Cbl-b also develop a severe autoimmune phenotype. In these mice, T cell hyperactivity is even more pronounced that in the single knockout and the process of ligand-induced TCR down-modulation is severely impaired [61].

Functions in peripheral tolerance

Studies on Cbl-b-deficient mice clearly suggest that Cbl-b plays a key role in peripheral tolerance. Cbl-b-deficient mice develop autoimmunity and their T cells are resistant to anergy and do not become unresponsive when receiving tolerogenic stimuli *in vivo* or *in vitro* [29, 52]. These effects can be explained by the fact that T cells from Cbl-b-deficient mice show reduced inactivation of PLC-γ1 in response to anergizing stimuli [29, 52]. In these cells, the kinetics of synapse disintegration are slower than in control anergized T cells, allowing T cell activation in the context of a more stable synapse [29]. In addition to its essential role in T cell anergy, Cbl-b has also been proposed to regulate B cell anergy [62] and suppression by regulatory T cells, as Cbl-b-deficient T cells are less sensitive to regulatory T cell-mediated suppression [63].

Itch

The E3 ligase Itch was first discovered in studies on the itchy mice, a spontaneous mutation characterized by coat-color alterations, constant itching of the skin, and the development of autoimmune disease. Genetic analysis identified a mutation in the *agouti* locus (a^{18H} in chromosome 2) that led to the disruption of the gene encoding the Itch E3 ligase [64].

Structure and protein interactions

Itch, which localizes to the endocytic compartment, contains a HECT domain that confers the E3 ligase activity, a C-terminal C2 domain (PKC-related C2 domain) and four WW domains that comprise two tryptophan residues separated by 20–22 amino acids. The C2 domains promote targeting of Itch to endosomes, while the WW domains bind with low affinity to proline-rich regions (Fig. 3). Itch has been shown to target members of the Jun family of transcription factors (i.e., JunB and c-Jun) for ubiquitination and subsequent proteasomal degradation in response to T cell activation [65]. Importantly, Jun proteins form part of AP-1 transcriptional complexes that play a crucial role in the transcription of several genes expressed during an immune response, such as IL-2 [24]. Itch has also been shown to ubiquitinate and promote the proteasome-mediated degradation of p63 and p73 [66, 67].

Regulation

Similar to Cbl-b, Itch expression is also up-regulated during anergy induction. T cells that are suboptimally stimulated induce the Ca^{2+}/calcineurin/NFAT-dependent expression of Itch [29]. The activity of Itch is also regulated by phosphorylation: JNK1-mediated phosphorylation induces a conformational change that activates Itch [68], whereas tyrosine phosphorylation by Fyn impairs the binding of Itch to one of its substrates, the transcription factor JunB [69]. As described for others E3 ligases, autoubiquitination can occur in Itch and the recruitment of the deubiquitinating enzyme FAM/USP9X may regulate this process [70].

Itch deficiency

Itchy mice, named this way because of the constant itching due to inflammation of the skin that results in chronic scarring, develop a systematic lymphoproliferative disorder characterized by the enlargement of peripheral lymphoid organs and chronic inflammation [64]. Itch-deficient T cells show also a biased differentiation towards T helper type 2 cells (Th2) and increase levels of Th2 cytokines (i.e., IL-4) and serum concentrations of IgG1 and IgE [71].

Functions in peripheral tolerance

The proteolytic degradation of specific signaling proteins has been studied in the context of the up-regulation of E3 ligases during anergy induction. When anergic T cells are re-stimulated, Itch localizes in detergent-insoluble membrane fractions, where it can target signaling molecules for degradation (Fig. 3). PLC-γ1, PKC-θ and RasGAP contain C2 domains, which may be involved in the interaction with Itch and the related E3 ligase Nedd4 [29]. Itch is required for ubiquitination of PLC-γ1 and PKC-θ, which seems to lead to sorting of these proteins into the endocytic pathway and eventual degradation by the lysosomes. Tsg101, a critical component of the ESCRT-1 complex required for the ubiquitin-dependent sorting to internal vesicles, may mediate this process [29]. A similar mechanism of action has been described for AIP4, the human homolog of Itch, which monoubiquitinates ligand-activated CXCR4 and sorts it to the lysosomal compartment through interactions with Hrs and Vsp4 proteins [72]. Similar to Cbl-b, Itch-deficient T cells show impaired induction of anergy [29]. A recent study has suggested that Itch may also interact with Notch in the regulation of T cell function, as a mouse deficient in Itch that overexpresses an active form of Notch in thymocytes develops an autoimmune phenotype of earlier onset than the one observed in itchy mice [73, 74]. A growing body of evidence supports the role of Notch signaling in the regulation of peripheral T cell activation and tolerance and Itch may also be involved in the regulation of this pathway.

GRAIL

GRAIL, a RING E3 transmembrane glycoprotein that localizes to vesicular structures in the cell, was first characterized in anergized T cells, where the mRNA and protein levels of this E3 ligase were markedly increased [75].

Structure and protein interactions

GRAIL contains an N-terminal signal peptide and a single transmembrane-spanning domain that promotes the localization of GRAIL in association with the endosomal compartment. This E3 ligase also contains a RING domain, a protease-associated (PA) domain and a coiled-coil region, which can bind the ubiquitin isopeptidase Otubain1. The C-terminal region of GRAIL can associate to different E2 enzymes to induce ubiquitination of target proteins.

Regulation

Otubain1 is a major regulator of GRAIL protein stability. This ubiquitin isopeptidase of the ovarian tumor family has two alternatively spliced isoforms with opposite functions. Otubain1 binds GRAIL and promotes its autoubiquitination and degradation by the proteasome. On the other hand, expression of Otubain 1 from an alternative reading frame (ARF-1) stabilizes the GRAIL protein, helped by the recruitment of the ubiquitin-specific protease 8 (USP-8) that promotes GRAIL deubiquitination [76].

GRAIL deficiency

Although a GRAIL knockout mouse has not been reported yet, generation of bone marrow chimeras in mice with cells expressing an enzymatically inactive dominant negative form of GRAIL has shown that this E3 ligase is required for the generation of peripheral T cell tolerance *in vivo* [77]. Similarly, overexpression of Otubain1, which induces the degradation of GRAIL, promotes resistance to the induction of T cell anergy [78].

Functions in peripheral tolerance

GRAIL plays a key role in the induction of anergy. In response to anergizing stimuli T cells activate the expression of GRAIL to induce a state of functional unresponsiveness. Up-regulation of GRAIL can be induced also with calcium ionophores and blocked with the calcineurin inhibitor cyclosporine A, which suggest the involvement of NFAT proteins in the regulation of GRAIL transcription in anergic T cells [29, 75]. Retroviral overexpression of GRAIL blocks IL-2 and IL-4 production and inhibits T cell proliferation [75]. Recent studies have attempted to characterize the mechanism by which GRAIL promotes the inhibition of IL-2 production and T cell proliferation. These studies have identified the Rho guanine dissociation inhibitor, RhoGDI, as a target substrate for ubiquitination by GRAIL (Fig. 3). Ubiquitination stabilizes this inhibitor, which results in deficient activation of Rho GTPases, preventing their effects on the reorganization of the cytoskeleton [78].

GRAIL expression is also increased in naturally arising regulatory T cells. Interestingly, overexpression of GRAIL in T cells seems to induce a suppressor phenotype in a T cell line. These experiments suggest that GRAIL may also be involved in the control of the generation and function of natural and peripherally induced regulatory T cells [79]. Similar to Itch, GRAIL may also form part of the Notch-induced program of regulation of T cell activation. Recently, Jagged-1-mediated activation of Notch signaling has been shown to inhibit human T cell proliferation and cytokine production while inducing up-regulation of the expression of GRAIL [80].

Concluding remarks

Significant progress has been made in the characterization of the mechanisms that regulate T cell anergy. The identification of several E3 ubiquitin ligases as effectors in the induction of T cell hyporesponsiveness has identified ubiquitination as a key process to prevent autoimmunity and induce tolerance. New exciting data have shed light on how this process can direct the localization and modulate the turnover of several key proteins, interfering directly with the signaling initiated at the immune synapse. However, our understanding of this complex mechanism is still in its early steps. The specific proteins targeted by each E3 ligase and the functional consequences of their ubiquitination remain to be uncovered. This information should prove beneficial for the development of therapeutic strategies able to modulate T cell tolerance in the treatment of autoimmune diseases and in the prevention of allograft rejection.

References

1 Palmer E (2003) Negative selection – clearing out the bad apples from the T-cell repertoire. *Nat Rev Immunol* 3: 383–391

2 Abbas AK, Lohr J, Knoechel B, Nagabhushanam V (2004) T cell tolerance and autoimmunity. *Autoimmun Rev* 3: 471–475

3 Schwartz RH (2003) T cell anergy. *Annu Rev Immunol* 21: 305–334

4 Macian F, Im SH, Garcia-Cozar FJ, Rao A (2004) T-cell anergy. *Curr Opin Immunol* 16: 209–216

5 Appleman LJ, Boussiotis VA (2003) T cell anergy and costimulation. *Immunol Rev* 192: 161–180

6 Greenwald RJ, Freeman GJ, Sharpe AH (2005) The B7 family revisited. *Annu Rev Immunol* 23: 515–548

7 Acuto O, Mise-Omata S, Mangino G, Michel F (2003) Molecular modifiers of T cell antigen receptor triggering threshold: The mechanism of CD28 costimulatory receptor. *Immunol Rev* 192: 21–31

8 Krogsgaard M, Huppa JB, Purbhoo MA, Davis MM (2003) Linking molecular and cellular events in T-cell activation and synapse formation. *Semin Immunol* 15: 307–315

9 Dustin ML (2005) A dynamic view of the immunological synapse. *Semin Immunol* 17: 400–410

10 Wulfing C, Davis MM (1998) A receptor/cytoskeletal movement triggered by costimulation during T cell activation. *Science* 282: 2266–2269

11 Jenkins MK, Schwartz RH (1987) Antigen presentation by chemically modified splenocytes induces antigen-specific T cell unresponsiveness *in vitro* and *in vivo*. *J Exp Med* 165: 302–319

12 Quill H, Schwartz RH (1987) Stimulation of normal inducer T cell clones with antigen presented by purified Ia molecules in planar lipid membranes: Specific induction of a long-lived state of proliferative nonresponsiveness. *J Immunol* 138: 3704–3712

13 Jenkins MK, Chen CA, Jung G, Mueller DL, Schwartz RH (1990) Inhibition of antigen-specific proliferation of type 1 murine T cell clones after stimulation with immobilized anti-CD3 monoclonal antibody. *J Immunol* 144: 16–22

14 Harding FA, McArthur JG, Gross JA, Raulet DH, Allison JP (1992) CD28-mediated signalling co-stimulates murine T cells and prevents induction of anergy in T-cell clones. *Nature* 356: 607–609

15 Egen JG, Kuhns MS, Allison JP (2002) CTLA-4: New insights into its biological function and use in tumor immunotherapy. *Nat Immunol* 3: 611–618

16 Waterhouse P, Penninger JM, Timms E, Wakeham A, Shahinian A, Lee KP, Thompson CB, Griesser H, Mak TW (1995) Lymphoproliferative disorders with early lethality in mice deficient in Ctla-4. *Science* 270: 985–988

17 Krummel MF, Allison JP (1995) CD28 and CTLA-4 have opposing effects on the response of T cells to stimulation. *J Exp Med* 182: 459–465

18 Perez VL, Van Parijs L, Biuckians A, Zheng XX, Strom TB, Abbas AK (1997) Induction of peripheral T cell tolerance *in vivo* requires CTLA-4 engagement. *Immunity* 6: 411–417

19 Greenwald RJ, Boussiotis VA, Lorsbach RB, Abbas AK, Sharpe AH (2001) CTLA-4 regulates induction of anergy *in vivo*. *Immunity* 14: 145–155

20 Rao A, Luo C, Hogan PG (1997) Transcription factors of the NFAT family: Regulation and function. *Annu Rev Immunol* 15: 707–747

21 Hogan PG, Chen L, Nardone J, Rao A (2003) Transcriptional regulation by calcium, calcineurin, and NFAT. *Genes Dev* 17: 2205–2232

22 Macian F (2005) NFAT proteins: Key regulators of T-cell development and function. *Nat Rev Immunol* 5: 472–484

23 Jain J, McCaffrey PG, Miner Z, Kerppola TK, Lambert JN, Verdine GL, Curran T, Rao A (1993) The T-cell transcription factor NFATp is a substrate for calcineurin and interacts with Fos and Jun. *Nature* 365: 352–355

24 Macian F, Lopez-Rodriguez C, Rao A (2001) Partners in transcription: NFAT and AP-1. *Oncogene* 20: 2476–2489

25 Macian F, Garcia-Cozar F, Im SH, Horton HF, Byrne MC, Rao A (2002) Transcriptional mechanisms underlying lymphocyte tolerance. *Cell* 109: 719–731

26 Heissmeyer V, Macian F, Varma R, Im SH, Garcia-Cozar F, Horton HF, Byrne MC, Feske S, Venuprasad K, Gu H et al (2005) A molecular dissection of lymphocyte

unresponsiveness induced by sustained calcium signalling. *Novartis Found Symp* 267: 165–174; discussion 174–179

27 Bandyopadhyay S, Soto-Nieves N, Macian F (2007) Transcriptional regulation of T cell tolerance. *Semin Immunol* 19: 180–187

28 Telander DG, Malvey EN, Mueller DL (1999) Evidence for repression of IL-2 gene activation in anergic T cells. *J Immunol* 162: 1460–1465

29 Heissmeyer V, Macian F, Im SH, Varma R, Feske S, Venuprasad K, Gu H, Liu YC, Dustin ML, Rao A (2004) Calcineurin imposes T cell unresponsiveness through targeted proteolysis of signaling proteins. *Nat Immunol* 5: 255–265

30 Zha Y, Marks R, Ho AW, Peterson AC, Janardhan S, Brown I, Praveen K, Stang S, Stone JC, Gajewski TF (2006) T cell anergy is reversed by active Ras and is regulated by diacylglycerol kinase-alpha. *Nat Immunol* 7: 1166–1173

31 Olenchock BA, Guo R, Carpenter JH, Jordan M, Topham MK, Koretzky GA, Zhong XP (2006) Disruption of diacylglycerol metabolism impairs the induction of T cell anergy. *Nat Immunol* 7: 1174–1181

32 Li L, Iwamoto Y, Berezovskaya A, Boussiotis VA (2006) A pathway regulated by cell cycle inhibitor p27(Kip1) and checkpoint inhibitor Smad3 is involved in the induction of T cell tolerance. *Nat Immunol* 7: 1157–1165

33 Boussiotis VA, Freeman GJ, Taylor PA, Berezovskaya A, Grass I, Blazar BR, Nadler LM (2000) p27kip1 functions as an anergy factor inhibiting interleukin 2 transcription and clonal expansion of alloreactive human and mouse helper T lymphocytes. *Nat Med* 6: 290–297

34 Powell JD, Lerner CG, Ewoldt GR, Schwartz RH (1999) The -180 site of the IL-2 promoter is the target of CREB/CREM binding in T cell anergy. *J Immunol* 163: 6631–6639

35 Bandyopadhyay S, Dure M, Paroder M, Soto-Nieves N, Puga I, Macian F (2007) Interleukin 2 gene transcription is regulated by Ikaros-induced changes in histone acetylation in anergic T cells. *Blood* 109: 2878–2886

36 Hundt M, Tabata H, Jeon MS, Hayashi K, Tanaka Y, Krishna R, De Giorgio L, Liu YC, Fukata M, Altman A (2006) Impaired activation and localization of LAT in anergic T cells as a consequence of a selective palmitoylation defect. *Immunity* 24: 513–522

37 Hochstrasser M (2006) Lingering mysteries of ubiquitin-chain assembly. *Cell* 124: 27–34

38 Fang S, Weissman AM (2004) A field guide to ubiquitylation. *Cell Mol Life Sci* 61: 1546–1561

39 Ciechanover A (2005) Proteolysis:from the lysosome to ubiquitin and the proteasome. *Nat Rev Mol Cell Biol* 6: 79–87

40 Schartner JM, Fathman CG, Seroogy CM (2007) Preservation of self: An overview of E3 ubiquitin ligases and T cell tolerance. *Semin Immunol* 19: 188–196

41 Harper JW, Schulman BA (2006) Structural complexity in ubiquitin recognition. *Cell* 124: 1133–1136

42 Nijman SM, Luna-Vargas MP, Velds A, Brummelkamp TR, Dirac AM, Sixma TK, Ber-

nards R (2005) A genomic and functional inventory of deubiquitinating enzymes. *Cell* 123: 773–786

43 Pickart CM, Fushman D (2004) Polyubiquitin chains: Polymeric protein signals. *Curr Opin Chem Biol* 8: 610–616

44 Ciechanover A, Heller H, Elias S, Haas AL, Hershko A (1980) ATP-dependent conjugation of reticulocyte proteins with the polypeptide required for protein degradation. *Proc Natl Acad Sci USA* 77: 1365–1368

45 Hershko A, Ciechanover A, Heller H, Haas AL, Rose IA (1980) Proposed role of ATP in protein breakdown:conjugation of protein with multiple chains of the polypeptide of ATP-dependent proteolysis. *Proc Natl Acad Sci USA* 77: 1783–1786

46 Ciechanover A, Elias S, Heller H, Ferber S, Hershko A (1980) Characterization of the heat-stable polypeptide of the ATP-dependent proteolytic system from reticulocytes. *J Biol Chem* 255: 7525–7528

47 Mukhopadhyay D, Riezman H (2007) Proteasome-independent functions of ubiquitin in endocytosis and signaling. *Science* 315: 201–205

48 Chen ZJ (2005) Ubiquitin signalling in the NF-kappaB pathway. *Nat Cell Biol* 7: 758–765

49 Wang HY, Altman Y, Fang D, Elly C, Dai Y, Shao Y, Liu YC (2001) Cbl promotes ubiquitination of the T cell receptor zeta through an adaptor function of Zap-70. *J Biol Chem* 276: 26004–26011

50 Chiang YJ, Kole HK, Brown K, Naramura M, Fukuhara S, Hu RJ, Jang IK, Gutkind JS, Shevach E, Gu H (2000) Cbl-b regulates the CD28 dependence of T-cell activation. *Nature* 403: 216–220

51 Krawczyk C, Bachmaier K, Sasaki T, Jones RG, Snapper SB, Bouchard D, Kozieradzki I, Ohashi PS, Alt FW, Penninger JM (2000) Cbl-b is a negative regulator of receptor clustering and raft aggregation in T cells. *Immunity* 13: 463–473

52 Jeon M-S, Atfield A, Venuprasad K, Krawczyk C, Sarao R, Elly C, Yang C, Arya S, Bachmaier K, Su L (2004) Essential Role of the E3 Ubiquitin ligase Cbl-b in T cell anergy induction. *Immunity* 21: 167–177

53 Wiedemann A, Muller S, Favier B, Penna D, Guiraud M, Delmas C, Champagne E, Valitutti S (2005) T-cell activation is accompanied by an ubiquitination process occurring at the immunological synapse. *Immunol Lett* 98: 57–61

54 Fang D, Liu Y-C (2001) Proteolysis-independent regulation of PI3K by Cbl-b-mediated ubiquitination in T cells. *Nat Immunol* 2: 870–875

55 Bustelo XR, Crespo P, Lopez-Barahona M, Gutkind JS, Barbacid M (1997) Cbl-b, a member of the Sli-1/c-Cbl protein family, inhibits Vav-mediated c-Jun N-terminal kinase activation. *Oncogene* 15: 2511–2520

56 Safford M, Collins S, Lutz MA, Allen A, Huang C-T, Kowalski J, Blackford A, Horton MR, Drake C, Schwartz RH et al (2005) Egr-2 and Egr-3 are negative regulators of T cell activation. *Nat Immunol* 6: 472–480

57 Zhang J, Bardos T, Li D, Gal I, Vermes C, Xu J, Mikecz K, Finnegan A, Lipkowitz

S, Glant TT (2002) Cutting edge: Regulation of T cell activation threshold by CD28 costimulation through targeting Cbl-b for ubiquitination. *J Immunol* 169: 2236–2240

58 Li D, Gal I, Vermes C, Alegre ML, Chong AS, Chen L, Shao Q, Adarichev V, Xu X, Koreny T et al (2004) Cutting edge: Cbl-b:one of the key molecules tuning CD28– and CTLA-4–mediated T cell costimulation. *J Immunol* 173: 7135–7139

59 Elly C, Witte S, Zhang Z, Rosnet O, Lipkowitz S, Altman A, Liu YC (1999) Tyrosine phosphorylation and complex formation of Cbl-b upon T cell receptor stimulation. *Oncogene* 18: 1147–1156

60 Bachmaier K, Krawczyk C, Kozieradzki I, Kong YY, Sasaki T, Oliveira-dos-Santos A, Mariathasan S, Bouchard D, Wakeham A, Itie A et al (2000) Negative regulation of lymphocyte activation and autoimmunity by the molecular adaptor Cbl-b. *Nature* 403: 211–216

61 Naramura M, Jang IK, Kole H, Huang F, Haines D, Gu H (2002) c-Cbl and Cbl-b regulate T cell responsiveness by promoting ligand-induced TCR down-modulation. *Nat Immunol* 3: 1192–1199

62 Kitaura Y, Jang IK, Wang Y, Han YC, Inazu T, Cadera EJ, Schlissel M, Hardy RR, Gu H (2007) Control of the B cell-intrinsic tolerance programs by ubiquitin ligases Cbl and Cbl-b. *Immunity* 26: 567–578

63 Wohlfert EA, Callahan MK, Clark RB (2004) Resistance to $CD4^+CD25^+$ regulatory T cells and TGF-β in Cbl-$b^{-/-}$ mice. *J Immunol* 173: 1059–1065

64 Perry WL, Hustad CM, Swing DA, O'Sullivan TN, Jenkins NA, Copeland NG (1998) The itchy locus encodes a novel ubiquitin protein ligase that is disrupted in a18H mice. *Nat Genet* 18: 143–146

65 Gao M, Labuda T, Xia Y, Gallagher E, Fang D, Liu YC, Karin M (2004) Jun turnover is controlled through JNK-dependent phosphorylation of the E3 ligase Itch. *Science* 306: 271–275

66 Rossi M, Aqeilan RI, Neale M, Candi E, Salomoni P, Knight RA, Croce CM, Melino G (2006) The E3 ubiquitin ligase Itch controls the protein stability of p63. *Proc Natl Acad Sci USA* 103: 12753–12758

67 Rossi M, De Laurenzi V, Munarriz E, Green DR, Liu YC, Vousden KH, Cesareni G, Melino G (2005) The ubiquitin-protein ligase Itch regulates p73 stability. *EMBO J* 24: 836–848

68 Gallagher E, Gao M, Liu YC, Karin M (2006) Activation of the E3 ubiquitin ligase Itch through a phosphorylation-induced conformational change. *Proc Natl Acad Sci USA* 103: 1717–1722

69 Yang C, Zhou W, Jeon MS, Demydenko D, Harada Y, Zhou H, Liu YC (2006) Negative regulation of the E3 ubiquitin ligase itch *via* Fyn-mediated tyrosine phosphorylation. *Mol Cell* 21: 135–141

70 Mouchantaf R, Azakir BA, McPherson PS, Millard SM, Wood SA, Angers A (2006) The ubiquitin ligase Itch is auto-ubiquitylated *in vivo* and *in vitro* but is protected from degradation by interacting with the deubiquitylating enzyme FAM/USP9X. *J Biol Chem* 281: 38738–38747

71 Fang D, Elly C, Gao B, Fang N, Altman Y, Joazeiro C, Hunter T, Copeland N, Jenkins N, Liu Y-C (2002) Dysregulation of T lymphocyte function in itchy mice: A role for Itch in TH2 differentiation. *Nat Immunol* 3: 281–287

72 Marchese A, Raiborg C, Santini F, Keen JH, Stenmark H, Benovic JL (2003) The E3 ubiquitin ligase AIP4 mediates ubiquitination and sorting of the G protein-coupled receptor CXCR4. *Dev Cell* 5: 709–722

73 Qiu L, Joazeiro C, Fang N, Wang HY, Elly C, Altman Y, Fang D, Hunter T, Liu YC (2000) Recognition and ubiquitination of Notch by Itch, a hect-type E3 ubiquitin ligase. *J Biol Chem* 275: 35734–35737

74 Matesic LE, Haines DC, Copeland NG, Jenkins NA (2006) Itch genetically interacts with Notch1 in a mouse autoimmune disease model. *Hum Mol Genet* 15: 3485–3497

75 Anandasabapathy N, Ford GS, Bloom D, Holness C, Paragas V, Seroogy C, Skrenta H, Hollenhorst M, Fathman CG, Soares L (2003) GRAIL: An E3 ubiquitin ligase that inhibits cytokine gene transcription is expressed in anergic $CD4^+$ T cells. *Immunity* 18: 535–547

76 Soares L, Seroogy C, Skrenta H, Anandasabapathy N, Lovelace P, Chung CD, Engleman E, Fathman CG (2004) Two isoforms of otubain 1 regulate T cell anergy *via* GRAIL. *Nat Immunol* 5: 45–54

77 Seroogy CM, Soares L, Ranheim EA, Su L, Holness C, Bloom D, Fathman CG (2004) The gene related to anergy in lymphocytes, an E3 ubiquitin ligase, is necessary for anergy induction in CD4 T cells. *J Immunol* 173: 79–85

78 Su L, Lineberry N, Huh Y, Soares L, Fathman CG (2006) A novel E3 ubiquitin ligase substrate screen identifies Rho guanine dissociation inhibitor as a substrate of gene related to anergy in lymphocytes. *J Immunol* 177: 7559–7566

79 MacKenzie DA, Schartner J, Lin J, Timmel A, Jennens-Clough M, Fathman CG, Seroogy CM (2007) GRAIL is up-regulated in $CD4^+$ $CD25^+$ T regulatory cells and is sufficient for conversion of T Cells to a regulatory phenotype. *J Biol Chem* 282: 9696–9702

80 Kostianovsky AM, Maier LM, Baecher-Allan C, Anderson AC, Anderson DE (2007) Up-regulation of gene related to anergy in lymphocytes is associated with notch-mediated human T cell suppression. *J Immunol* 178: 6158–6163

FOXP3 biochemistry will lead to novel drug approaches for vaccines and diseases that lack suppressor T cells

Bin Li, Xiaomin Song, Arabinda Samanta, Kathryn Bembas, Amy Brown, Geng Zhang, Makoto Katsumata, Yuan Shen, Sandra J. Saouaf and Mark I. Greene

Department of Pathology and Laboratory Medicine, University of Pennsylvania, 252 John Morgan, 36th and Hamilton Walk, Philadelphia, PA 19104-6082, USA

Introduction

FOXP3 is a forkhead family transcription factor, which acts as a master regulator in the development of natural regulatory T cells (Tregs) and their function in the control of self tolerance [1]. Natural Tregs, which represent about 5–10% of total $CD4^+$ T cells, develop in the thymus and have a middle-high TCR-binding affinity. Tregs function as suppressors of multiple immune cells including CD4 effector T cells, CD8 cytotoxic T cells, B cells, NK cells, and dendritic cells *in vivo* [2]. Although the molecular mechanism by which Tregs suppress these multiple immune cells in a cell-cell contact-dependent manner is largely unknown [3], recent experimental evidence supports the notion that the level and duration of FOXP3 expression is essential to Treg-mediated dominant suppression [4, 5]. A complete understanding of the biochemistry of FOXP3 activity in Tregs will have therapeutic implications for transplantation, allergy, autoimmune disease, inflammatory disease, vaccine development and cancer [6].

The brief history of Tregs

Nearly 38 years ago, Sakaura and Nishizuka published a study showing that neonatal thymectomy caused oophoritis in mice, suggesting that cells developed in thymus could suppress autoimmunity [7]. In the 1970s studies from Gershon and Kondo [8, 9] indicated that these suppressive cells might be T cells by demonstrating that T cells suppressed antibody responses. The term "suppressor T cells" was first introduced in the review of Gershon et al. in 1972 [10]. The studies on suppressor T cells included $CD8^+$ T cells, which target the major histocompatibility complex (MHC) class I molecule Qa-1 protein [11, 12]. During the 1980s, the field of suppressor T cells was itself suppressed due to the unclear molecular character of the immunosuppressive factors. The phenotypic and functional identification of these suppressor T cells was advanced by the depletion study of a subpopulation of T cells

The Immune Synapse as a Novel Target for Therapy, edited by Luis Graca

[13]. The first experimental evidence to support the notion that these suppressor T cells could be a $CD4^+$ T cell subset came from Powrie, Mason and colleagues [14, 15]. Around the same period of time, Foxp3 was identified as the responsive gene for X-linked fatal lymphoreticular disease in the scurfy mice, and $CD4^+$ T cells were subsequently found as the mediator of the disease [16–20]. In 1993, Waldmann and colleagues [21] published an important study of infectious transplantation tolerance, in which naïve cells acquire a tolerant phenotype following their coexistence with tolerant cells. In 1995, Sakaguchi and colleagues [22] identified $CD4^+CD25^+$ T cells as the subpopulation of T cells responsible for immune regulation, and the field of suppressor T cells was reborn. Since then, numerous subpopulations of T cells have been demonstrated to contain Treg activity [23]. *In vitro* characterization of Treg function demonstrated their inhibition of IL-2 production [24]; however, the *in vivo* dynamics of Treg-mediated suppression is less clear and less consistent with function *in vitro* [25].

As a transcriptional repressor of cytokine gene expression [26], FOXP3 was subsequently identified as the "essential and sufficient" transcription factor for the development and function of natural $CD4^+CD25^+$ Tregs by three independent studies [27–29]. The mutated human *FOXP3* gene was also identified as being responsible for the human disorder, X-linked autoimmunity–allergic dysregulation syndrome (XLAAD) [30]. Our laboratory has been interested in the biochemical features of these suppressor T cells and their function [31–39]. Currently, our focus is on the mechanism by which the FOXP3 ensemble, as a transcriptional regulatory complex, mediates the suppressive phenotype in Tregs.

Modulation of FOXP3 level and Treg function

Various cytokines, bacterial pathogens, monoclonal antibodies targeting cell surface receptors and pharmaceutical small molecular agents have been noted as modulators of Treg function in mice and humans. For instance, Pasare and Medzhitov [40] found that the inflammatory cytokine IL-6 relieves $CD4^+CD25^+$ Treg-mediated suppression of effector T cells upon microbial induction of the Toll-like receptor pathway on dendritic cells. Soluble TNF (50 ng/mL) and anti-TNFRII agonistic monoclonal antibody treatments lower the level of FOXP3 transcription and expression such that $CD4^+CD25^{hi}$ Treg function is downmodulated in humans [41]. Moreover, bacterial lipopolysaccharide (LPS) activates $CD4^+CD25^+$ Treg cells and down-regulates FOXP3 expression in a time-dependent manner [42, 43]. However, Chen et al. [44] found that immunosuppressive cytokine TGF-β plus TCR stimulation induced FOXP3 expression in $CD4^+CD25^-$ effector T cells with suppressive function *in vitro* and *in vivo*. Similarly, IFN-γ treatment increased the transcription and expression of Foxp3 and enhanced Treg function [45]. *In vitro* treatment and administration of the peptide copolymer-I (COP-I) induced Foxp3 expression and promoted the conversion of peripheral $CD4^+CD25^-$ effector T cells to $CD4^+CD25^+$ Tregs in human and mouse cells [46].

The role of $CD4^+CD25^+$ Tregs in controlling inflammation has been well documented in recent studies [47, 48]. It is not surprising that more agents are being reported as modulators of Treg function through up- or down- regulation of FOXP3 levels. Understanding how these molecules affect FOXP3 expression and ensembles in Tregs is fundamental for developing novel drug approaches to control inflammatory immune diseases.

Molecular biology and biochemistry of the FOXP3 ensemble in T cell regulation

The Forkhead box P subfamily transcription factors include FOXP1, FOXP2, FOXP3 and FOXP4. Both FOXP1 and FOXP3 are expressed in $CD4^+CD25^+$ T Tregs, and all of the FOXP subfamily members share the conserved Zinc finger-leucine zipper tetramerization domain and forkhead DNA binding domain [49]. The N-terminal sequence of FOXP3 is proline rich, and has very little similarity with the N-terminal sequences of FOXP1, FOXP2 and FOXP4, which are glutamine rich. Moreover, the C terminus of FOXP1, FOXP2 and FOXP4 contains an extra 132 amino acids following the forkhead DNA binding domain that are lacking in the C terminus of FOXP3. FOXP3 was previously found as a transcriptional repressor of cytokine genes, such as IL-2, after TCR stimulation [26]. Both the leucine zipper domain and the C-terminal forkhead domain are required for FOXP3-repressive function [49–51]. The conserved sequence recognized by the forkhead DNA binding domain of FOXP3 targeted genes was recently identified [52]. Interestingly, many of these FOXP3 targeted genes are critical modulators of T cell activation and function [52]. Moreover, genes directly targeted by FOXP3 in mouse $CD4^+Foxp3^+$ Tregs were found to be either up-regulated or down-regulated, suggesting that Foxp3 may function as both a transcriptional repressor and activator [53].

The forkhead domain of FOXP subfamily protein appears to interact with NFATc2 (nuclear factor of activated T cells, cytoplasmic, calcineurin-dependent 2), which may contribute to its function as a passive transcriptional repressor [54, 55]. The Zinc finger and leucine zipper domains of FOXP3 mediate its homoassociation [50, 51], homotetramerization [49] and heteromerization with subfamily member FOXP1 [49, 56]. The C-terminal linker region between the tetramerization domain and the forkhead DNA binding domain of FOXP3 directly interacts with the C-terminal transcriptional repression domain of AML family transcription factor AML1 (acute myeloid leukemia 1)/Runx1 (Runt-related transcription factor 1), and detailed mutation analysis suggests that this interaction is also critical for FOXP3-mediated suppression of IL-2 production after T cell activation [57]. The N-terminal exon 2 encoding region, which lacks the smaller splicing isoform of human FOXP3, was found to be associated with orphan retinoic acid nuclear receptor RORa [58]. However, the molecular mechanisms that drive NFAT, AML1 and FOXP3 interaction as well as the effect of these interactions on Treg function are still unclear.

The unique features of FOXP3 function in Tregs may be dependent on its N-terminal proline-rich domain and its complex formation with other transcriptional co-regulatory proteins. Recently, a critical mechanism by which FOXP3 actively represses transcription by this N-terminal proline-rich domain was identified [33]. The FOXP3 N-terminal proline rich domain was required to mediate FOXP3 repressive activity and association with the histone acetyltransferase (HAT) TIP60 and class II histone deacetylases (HDAC), including HDAC7 and HDAC9 in human $CD4^{+}CD25^{+}$ Tregs, thus forming the FOXP3 complex ensemble [33, 59].

These histone modification enzymes may be recruited by FOXP3 to chromatin through its forkhead DNA binding domain, or through other FOXP3 binding partners having DNA binding capabilities, such as transcription factor NFAT or AML1, to modify site-specific gene transcription in Tregs [60]. HATs also modify non-histone proteins, and FOXP3 was found as an acetylated protein in Tregs [33]. The acetylation status may also affect the FOXP3 complex ensemble since association of HDAC9 with the FOXP3 complex is lost after TCR stimulation [33]. Treatment of the cells with histone deacetylation inhibitor trichostatin A restored HDAC9 association [33].

Our efforts to identify other molecules belonging to the FOXP3 complex ensemble in T cells using MS/Qstar spectrophotometric analysis of immunoprecipitated FOXP3 from nuclear extracts of FLAG-FOXP3 ectopically expressing T cells (Li and Greene, unpublished observations) identified several ATP-dependent chromatin-remodeling enzymes. Interestingly, these chromatin-remodeling factors co-fractionate precisely with FOXP3 in a large molecular weight complex by size-dependent gel filtration (Li and Greene, unpublished observations). This large molecular weight FOXP3 complex ensemble has a distribution pattern distinct from that of the lower molecular weight FOXP3 containing complex including other FOXP3-associated transcription factors such as FOXP1 and NFATc2, and histone modification enzymes such as TIP60 and HDAC7 (Li and Greene, unpublished observations). The physiological roles of these chromatin-remodeling enzymes in the FOXP3 complex ensemble remain unclear.

Conclusion

The complexity of the FOXP3 ensembles, which include other transcription factors and enzymes involved in histone modification as well as chromatin remodeling, may indicate that multiple types of regulation become imposed on Tregs. Signals from cytokines, receptors, etc., may affect regulation by modifying components of the FOXP3 ensemble, causing post-translational modifications of FOXP3 and associated proteins, and even influencing FOXP3 disposition within the Treg. Developing pharmaceutical agents specifically targeting the enzymatic activities of the FOXP3 complex ensemble represents a new therapeutic opportunity in vaccine development and therapy of immune diseases through modulating Treg function.

Reference

1 Sakaguchi S (2004) Naturally arising CD4+ regulatory T cells for immunologic self-tolerance and negative control of immune responses. *Annu Rev Immunol* 22: 531–562

2 Rudensky AY, Campbell DJ (2006) *In vivo* sites and cellular mechanisms of Treg cell-mediated suppression. *J Exp Med* 203: 489–492

3 von Boehmer H (2005) Mechanisms of suppression by suppressor T cells. *Nat Immunol* 6: 338–344

4 Williams LM, Rudensky AY (2007) Maintenance of the Foxp3-dependent developmental program in mature regulatory T cells requires continued expression of Foxp3. *Nat Immunol* 8: 277–284

5 Wan YY, Flavell RA (2007) Regulatory T-cell functions are subverted and converted owing to attenuated Foxp3 expression. *Nature* 445: 766–770

6 Li B, Samanta A, Song X, Furuuchi K, Iacono KT, Kennedy S, Katsumata M, Saouaf SJ, Greene MI (2006) FOXP3 ensembles in T-cell regulation. *Immunol Rev* 212: 99–113

7 Nishizuka Y, Sakakura T (1969) Thymus and reproduction: sex-linked dysgenesia of the gonad after neonatal thymectomy in mice. *Science* 166: 753–755

8 Gershon RK, Kondo K (1971) Infectious immunological tolerance. *Immunology* 21: 903–914

9 Gershon RK, Kondo K (1970) Cell interactions in the induction of tolerance: the role of thymic lymphocytes. *Immunology* 18: 723–737

10 Gershon RK, Cohen P, Hencin R, Liebhaber SA (1972) Suppressor T cells. *J Immunol* 108: 586–590

11 Cantor H, Hugenberger J, McVay-Boudreau L, Eardley DD, Kemp J, Shen FW, Gershon RK (1978) Immunoregulatory circuits among T-cell sets. Identification of a subpopulation of T-helper cells that induces feedback inhibition. *J Exp Med* 148: 871–877

12 Lu L, Werneck MB, Cantor H (2006) The immunoregulatory effects of Qa-1. *Immunol Rev* 212: 51–59

13 Sakaguchi S, Fukuma K, Kuribayashi K, Masuda T (1985) Organ-specific autoimmune diseases induced in mice by elimination of T cell subset. I. Evidence for the active participation of T cells in natural self-tolerance; deficit of a T cell subset as a possible cause of autoimmune disease. *J Exp Med* 161: 72–87

14 Fowell D, McKnight AJ, Powrie F, Dyke R, Mason D (1991) Subsets of CD4+ T cells and their roles in the induction and prevention of autoimmunity. *Immunol Rev* 123: 37–64

15 Powrie F, Mason D (1990) OX-22high CD4+ T cells induce wasting disease with multiple organ pathology: prevention by the OX-22low subset. *J Exp Med* 172: 1701–1708

16 Godfrey VL, Wilkinson JE, Russell LB (1991) X-linked lymphoreticular disease in the scurfy (sf) mutant mouse. *Am J Pathol* 138: 1379–1387

17 Blair PJ, Carpenter DA, Godfrey VL, Russell LB, Wilkinson JE, Rinchik EM (1994) The mouse scurfy (sf) mutation is tightly linked to Gata1 and Tfe3 on the proximal X chromosome. *Mamm Genome* 5: 652–654

18 Brunkow ME, Jeffery EW, Hjerrild KA, Paeper B, Clark LB, Yasayko SA, Wilkinson JE, Galas D, Ziegler SF, Ramsdell F (2001) Disruption of a new forkhead/winged-helix protein, scurfin, results in the fatal lymphoproliferative disorder of the scurfy mouse. *Nat Genet* 27: 68–73

19 Bennett CL, Christie J, Ramsdell F, Brunkow ME, Ferguson PJ, Whitesell L, Kelly TE, Saulsbury FT, Chance PF, Ochs HD (2001) The immune dysregulation, polyendocrinopathy, enteropathy, X-linked syndrome (IPEX) is caused by mutations of FOXP3. *Nat Genet* 27: 20–21

20 Wildin RS, Ramsdell F, Peake J, Faravelli F, Casanova JL, Buist N, Levy-Lahad E, Mazzella M, Goulet O, Perroni L et al (2001) X-linked neonatal diabetes mellitus, enteropathy and endocrinopathy syndrome is the human equivalent of mouse scurfy. *Nat Genet* 27: 18–20

21 Qin S, Cobbold SP, Pope H, Elliott J, Kioussis D, Davies J, Waldmann H (1993) "Infectious" transplantation tolerance. *Science* 259: 974–977

22 Takahashi T, Kuniyasu Y, Toda M, Sakaguchi N, Itoh M, Iwata M, Shimizu J, Sakaguchi S (1998) Immunologic self-tolerance maintained by $CD25^+CD4^+$ naturally anergic and suppressive T cells: induction of autoimmune disease by breaking their anergic/suppressive state. *Int Immunol* 10: 1969–1980

23 Shevach EM (2006) From vanilla to 28 flavors: multiple varieties of T regulatory cells. *Immunity* 25: 195–201

24 Thornton AM, Shevach EM (1998) $CD4^+CD25^+$ immunoregulatory T cells suppress polyclonal T cell activation *in vitro* by inhibiting interleukin 2 production. *J Exp Med* 188: 287–296

25 Klein L, Khazaie K, von Boehmer H (2003) *In vivo* dynamics of antigen-specific regulatory T cells not predicted from behavior *in vitro*. *Proc Natl Acad Sci USA* 100: 8886–8891

26 Schubert LA, Jeffery E, Zhang Y, Ramsdell F, Ziegler SF (2001) Scurfin (FOXP3) acts as a repressor of transcription and regulates T cell activation. *J Biol Chem* 276: 37672–37679

27 Khattri R, Cox T, Yasayko SA, Ramsdell F (2003) An essential role for Scurfin in $CD4^+CD25^+$ T regulatory cells. *Nat Immunol* 4: 337–342

28 Hori S, Nomura T, Sakaguchi S (2003) Control of regulatory T cell development by the transcription factor Foxp3. *Science* 299: 1057–1061

29 Fontenot JD, Gavin MA, Rudensky AY (2003) Foxp3 programs the development and function of $CD4^+CD25^+$ regulatory T cells. *Nat Immunol* 4: 330–336

30 Chatila TA, Blaeser F, Ho N, Lederman HM, Voulgaropoulos C, Helms C, Bowcock AM (2000) JM2, encoding a fork head-related protein, is mutated in X-linked autoimmunity-allergic disregulation syndrome. *J Clin Invest* 106: R75–81

31 Maeda H, Fujimoto S, Greene MI (2000) Suppressor T cells regulate the nonanergic cell population that remains after peripheral tolerance is induced to the Mls-1 antigen in T cell receptor Vbeta 8.1 transgenic mice. *Proc Natl Acad Sci USA* 97: 13257–13262

32 Saouaf SJ, Brennan PJ, Shen Y, Greene MI (2003) Mechanisms of peripheral immune tolerance: conversion of the immune to the unresponsive phenotype. *Immunol Res* 28: 193–199

33 Li B, Samanta A, Song X, Iacono KT, Bembas K, Tao R, Basu S, Riley JL, Hancock WW, Shen Y et al (2007) FOXP3 interactions with histone acetyltransferase and class II histone deacetylases are required for repression. *Proc Natl Acad Sci USA* 104: 4571–4576

34 Weiner DB, Williams WV, Siegel RM, Jerrold-Jones S, Greene MI (1988) Molecular characterization of suppressor T cells. *Transplant Proc* 20: 1151–1153

35 Schatten S, Granstein RD, Drebin JA, Greene MI (1984) Suppressor T cells and the immune response to tumors. *Crit Rev Immunol* 4: 335–379

36 Hirai Y, Dohi Y, Sy MS, Greene MI, Nisonoff A (1981) Suppressor T cells induced by idiotype-coupled cells function across an allotype barrier. *J Immunol* 126: 2064–2066

37 Bromberg JS, Benacerraf B, Greene MI (1981) Mechanisms of regulation of cell-mediated immunity. VII. Suppressor T cells induced by suboptimal doses of antigen plus an I-J-specific allogeneic effect. *J Exp Med* 153: 437–449

38 Greene MI, Bach BA, Benacerraf B (1979) Mechanisms of regulation of cell-mediated immunity. III. The characterization of azobenzenearsonate-specific suppressor T-cell-derived-suppressor factors. *J Exp Med* 149: 1069–1083

39 Greene MI, Fujimoto S, Sehon AH (1977) Regulation of the immune response to tumor antigens. III. Characterization of thymic suppressor factor(s) produced by tumor-bearing hosts. *J Immunol* 119: 757–764

40 Pasare C, Medzhitov R (2003) Toll pathway-dependent blockade of $CD4^+CD25^+$ T cell-mediated suppression by dendritic cells. *Science* 299: 1033–1036

41 Valencia X, Stephens G, Goldbach-Mansky R, Wilson M, Shevach EM, Lipsky PE (2006) TNF downmodulates the function of human $CD4^+CD25^{hi}$ T-regulatory cells. *Blood* 108: 253–261

42 Caramalho I, Lopes-Carvalho T, Ostler D, Zelenay S, Haury M, Demengeot J (2003) Regulatory T cells selectively express toll-like receptors and are activated by lipopolysaccharide. *J Exp Med* 197: 403–411

43 Lewkowicz P, Lewkowicz N, Sasiak A, Tchorzewski H (2006) Lipopolysaccharide-activated $CD4^+CD25^+$ T regulatory cells inhibit neutrophil function and promote their apoptosis and death. *J Immunol* 177: 7155–7163

44 Chen W, Jin W, Hardegen N, Lei KJ, Li L, Marinos N, McGrady G, Wahl SM (2003) Conversion of peripheral $CD4^+CD25^-$ naive T cells to $CD4^+CD25^+$ regulatory T cells by TGF-beta induction of transcription factor Foxp3. *J Exp Med* 198: 1875–1886

45 Wang Z, Hong J, Sun W, Xu G, Li N, Chen X, Liu A, Xu L, Sun B, Zhang JZ (2006) Role of IFN-gamma in induction of Foxp3 and conversion of $CD4^+$ $CD25^-$ T cells to $CD4^+$ Tregs. *J Clin Invest* 116: 2434–2441

46 Hong J, Li N, Zhang X, Zheng B, Zhang JZ (2005) Induction of $CD4^+CD25^+$ regulatory T cells by copolymer-I through activation of transcription factor Foxp3. *Proc Natl Acad Sci USA* 102: 6449–6454
47 Izcue A, Coombes JL, Powrie F (2006) Regulatory T cells suppress systemic and mucosal immune activation to control intestinal inflammation. *Immunol Rev* 212: 256–271
48 Read S, Malmstrom V, Powrie F (2000) Cytotoxic T lymphocyte-associated antigen 4 plays an essential role in the function of $CD25(^+)CD4(^+)$ regulatory cells that control intestinal inflammation. *J Exp Med* 192: 295–302
49 Li B, Samanta A, Song X, Iacono KT, Brennan PJ, Chatila TA, Roncador G, Banham AH, Riley JL, Wang Q et al (2007) FOXP3 is a homo-oligomer and a component of a supramolecular regulatory complex disabled in the human XLAAD/IPEX autoimmune syndrome. *Int Immunol* 19: 825–835
50 Chae WJ, Henegariu O, Lee SK, Bothwell AL (2006) The mutant leucine-zipper domain impairs both dimerization and suppressive function of Foxp3 in T cells. *Proc Natl Acad Sci USA* 103: 9631–9636
51 Lopes JE, Torgerson TR, Schubert LA, Anover SD, Ocheltree EL, Ochs HD, Ziegler SF (2006) Analysis of FOXP3 reveals multiple domains required for its function as a transcriptional repressor. *J Immunol* 177: 3133–3142
52 Marson A, Kretschmer K, Frampton GM, Jacobsen ES, Polansky JK, MacIsaac KD, Levine SS, Fraenkel E, von Boehmer H, Young RA (2007) Foxp3 occupancy and regulation of key target genes during T-cell stimulation. *Nature* 445: 931–935
53 Zheng Y, Josefowicz SZ, Kas A, Chu TT, Gavin MA, Rudensky AY (2007) Genome-wide analysis of Foxp3 target genes in developing and mature regulatory T cells. *Nature* 445: 936–940
54 Bettelli E, Dastrange M, Oukka M (2005) Foxp3 interacts with nuclear factor of activated T cells and NF-kappa B to repress cytokine gene expression and effector functions of T helper cells. *Proc Natl Acad Sci USA* 102: 5138–5143
55 Wu Y, Borde M, Heissmeyer V, Feuerer M, Lapan AD, Stroud JC, Bates DL, Guo L, Han A, Ziegler SF et al (2006) FOXP3 controls regulatory T cell function through cooperation with NFAT. *Cell* 126: 375–387
56 Wang B, Lin D, Li C, Tucker P (2003) Multiple domains define the expression and regulatory properties of Foxp1 forkhead transcriptional repressors. *J Biol Chem* 278: 24259–24268
57 Ono M, Yaguchi H, Ohkura N, Kitabayashi I, Nagamura Y, Nomura T, Miyachi Y, Tsukada T, Sakaguchi S (2007) Foxp3 controls regulatory T-cell function by interacting with AML1/Runx1. *Nature* 446: 685–689
58 Ziegler SF (2006) FOXP3: of mice and men. *Annu Rev Immunol* 24: 209–226
59 Li B, Greene MI (2007) FOXP3 actively represses transcription by recruiting the HAT/HDAC complex. *Cell Cycle* 6: 1432–1436
60 Li B, Saouaf SJ, Samanta A, Shen Y, Hancock WW, Greene MI (2007) Biochemistry and therapeutic implications of understanding mechanisms underlying FOXP3 activity. *Curr Opin Immunol,* DOI 10.1016/J.COI.2007.07.006

Transforming growth factor-β: From its effect in T cell activation to a role in dominant tolerance

Ramireddy Bommireddy[1,2] *and Thomas Doetschman*[1,3,4]

[1]BIO5 Institute, [2]Department of Immunobiology, [3]Department of Cell Biology and Anatomy, [4]Arizona Cancer Center, University of Arizona, PO Box 245217, Tucson, AZ 85724-5217, USA

Introduction

Transforming growth factor (TGF)β1 is an important immunoregulatory cytokine involved in maintenance of self tolerance and T cell homeostasis [1]. It is produced by several immune and non-immune cell types and functions in both autocrine and paracrine manners [1–3]. Regulatory T cells (Treg) cells are the primary source of TGFβ1 for controlling autoimmune disease. Blockade of TGFβ signaling in T cells causes severe autoimmune disease in mice [4, 5]. However, TGFβ signaling in T cells under inflammatory conditions also causes the development of experimentally induced autoimmune encephalomyelitis (EAE), a mouse model for human multiple sclerosis (MS) disease [6]. This review focuses on recent findings concerning the function of TGFβ1 in T cells and immune tolerance, and it discusses implications of these findings for therapeutic intervention in autoimmune and inflammatory diseases.

TGFβ signaling in T cells

TGFβ1 inhibits T cell response to antigenic stimulation *in vitro*, suggesting that these cells express receptors for TGFβ1. TGFβ1 signals primarily through membrane-bound serine/threonine kinase receptors, and its SMAD signaling intermediates vary depending on cell type and response (Fig. 1) [7, 8]. This pathway induces many genes for proteins such as $p21^{Cip1}$ (*Cdkn1a*), $p27^{kip1}$ (*Cdkn1b*), IL-2 receptor α chain or CD25 (*Il2ra*) and other molecules in T cells. CD25 is the α subunit of the IL-2 high-affinity receptor expressed on activated T cells and Treg cells. The addition of TGFβ1 to T cells along with anti-CD3 and anti-CD28 antibodies has been shown to induce *Foxp3* in $CD4^+CD25^-$ T cells and to induce proliferation of $CD4^+CD25^+$ Treg cells [9–11]. TGFβ1 induces *Ctla4* in naïve T cells, which in turn is responsible for the induction of *Foxp3* in inducible Treg cells [12].

The Immune Synapse as a Novel Target for Therapy, edited by Luis Graca

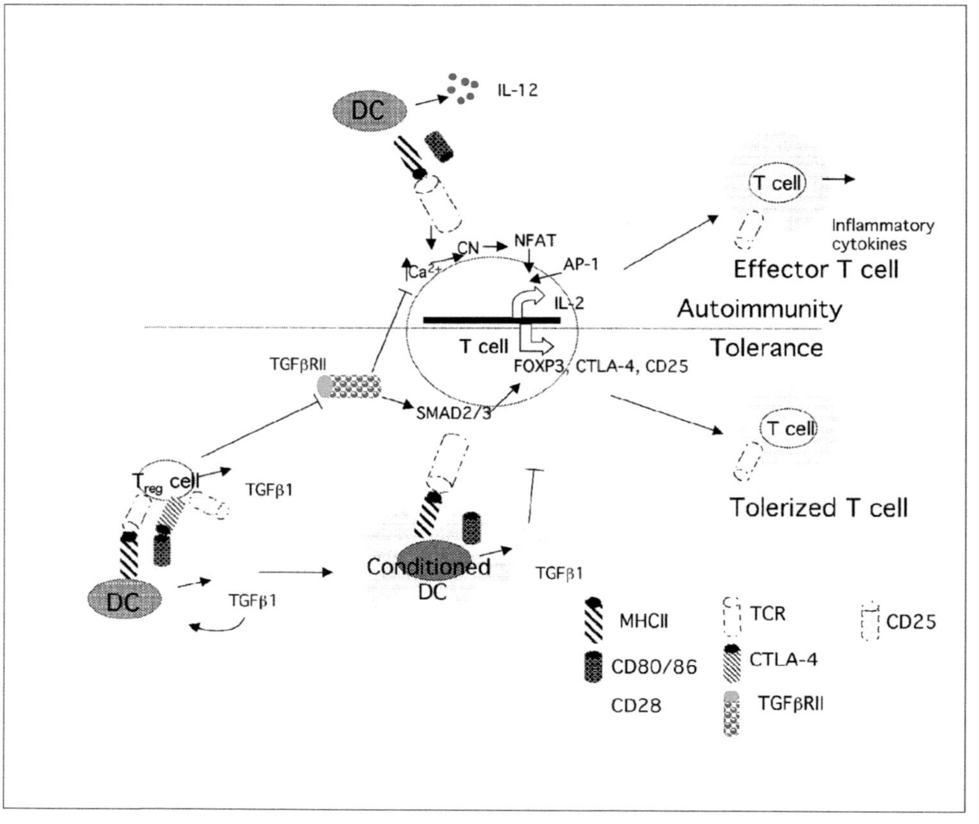

Figure 1

Antigen presentation by mature dendritic cells (DC; produce IL-12) to T cells leads to activation of T cells.

T cell activation is caused by elevation of cytosolic Ca^{2+} levels, which activate calcineurin (CN). CN activates NFAT. NFAT in the presence of AP-1 induces IL-2 and other immune response genes. Activated T cells produce inflammatory cytokines and cause autoimmunity (upper part of the signaling scheme) due to either increased costimulation or decreased TGFβ signaling. Regulatory T (Treg) cells interact with DC through CTLA-4/CD80 and TCR/MHC class II (MHCII) interactions. This process leads to conditioning of DC through production of TGFβ1. TGFβ1 is also produced by Treg cells and is present on the surface of Treg cells. TGFβ1 acts both in an autocrine and paracrine manner and causes down regulation of costimulatory molecules on DC. TGFβ1 causes inhibition of Ca^{2+} flux and T cell activation. TGFβ1 also induces Foxp3 *in T cells. Antigen presentation by immature DC or conditioned DC (DC that have interacted with Treg cells; produce TGFβ1) to T cells leads to induction of tolerance (lower part of the signaling scheme). Treg cells are generated during the tolerance process due to induction of* Foxp3 *in T cells by IL-2 and TGFβ1.*

TGFβ1 also signals through a SMAD-independent Ca^{2+}-calcineurin-NF-AT cascade, which inhibits naïve T cell activation [13, 14]. While addition of TGFβ1 to T cells causes inhibition of $[Ca^{2+}]_i$ flux, ablation of *Tgfb1 in vivo* causes elevation of $[Ca^{2+}]_i$ and increased flux and hyperresponsiveness to stimulation [13, 15]. TGFβ1 also induces *Ctla4* and *Foxp3* expression in naïve T cells and causes Th to Treg cell conversion and subsequent expansion through SMAD signaling (Fig. 1) [9, 12]. Conditional deletion in T cells of the TGFβ receptor type II gene (*Tgfbr2*) causes T cell resistance to Treg cell suppressor function, confirming that TGFβ signaling is essential for T cell homeostasis and Treg cell suppressor function [4, 5].

The growth inhibitory effects of TGFβ1 are thought to be mediated through the induction of genes for $p15^{INK4b}$ (*Cdkn2b*), *Cdkn1a* and *Cdkn1b* in a SMAD-independent manner [16, 17]. However, $p15^{INK4b}$ is dispensable for the growth inhibitory effect of TGFβ1, whereas the inhibition of *Cdk4* by TGFβ1 is SMAD3 dependent in T cells [17]. In the same study it was shown that activation-induced death of T cells is prevented by TGFβ1, irrespective of the presence or absence of *Cdkn1a* and *Cdkn1b*. This study also showed that TGFβ1 causes growth arrest of *Cdkn1a* and *Cdkn1b* double-knockout T cells when stimulated under low costimulatory conditions, suggesting an alternative inhibitory pathway for TGFβ1 in T cells. TGFβ1 has been shown to induce apoptosis through the inhibition of the Bcl-xl gene (*Bcl2l1*) and to promote cytochrome c release in epithelial cells [18]. However, it inhibits apoptosis of T cells in the presence of CD28 costimulation and enhances proliferation [19]. TGFβ1 induces survival factor genes such as *Bcl2*, *Bcl2l1* and growth inhibitory genes such as *Cdkn1a* in activated T cells [20, 21]. We have shown that $Tgfb1^{-/-}$ T cells are hyperresponsive when they are stimulated with suboptimal doses of mitogens [13, 14]. Addition of TGFβ1 neutralizing antibody to wild-type thymocyte cultures does not enhance their response, suggesting that this $Tgfb1^{-/-}$ thymocyte hyperresponsiveness is cell intrinsic.

TGFβ1 is known to inhibit differentiation of Th1 and Th2 cells, whereas it induces Th17 cell differentiation from naïve $CD4^+$ cells upon anti-CD3 and LPS stimulation in the presence of dendritic cells (DC). However, Th17 cells fail to develop from DN-TGFβRII $CD4^+$ T cells under the same culture conditions, suggesting that TGFβ signaling is required for Th17 cell differentiation. Nuclear orphan receptor RORγt is required for the development of Th17 cells. TGFβ1 and IL-6 induce RORγt independently. FOXP3, which is induced by TGFβ1, inhibits RORγt-induced IL-17 expression [22]. Thus, a balance of IL-6 and TGFβ1 determines whether a T cell becomes a Treg or a Th17 cell.

TGFβ1 inhibits both Th1 and Th2 cell differentiation by inhibiting T-bet (*Tbx21*) and *Gata3* expression, respectively [2, 23, 24]. TGFβ1 inhibits *Ifng* expression through inhibition of *Tbx21* and *Stat4*, which are known to induce Th1 cell differentiation [2]. Overexpression of *Tbx21* in T cells prevents the inhibitory effect of TGFβ1 on Th1 cell differentiation but does not prevent its inhibition of *Ifng* expression. Similarly, the overexpression of *Stat4* partially impairs TGFβ1 inhibition of *Ifng* but does not prevent the inhibition of Th1 differentiation. These data suggest that TGFβ1 uses distinct signaling mechanisms in T cells to inhibit *Ifng* and Th1 cell differentiation [23]. The inhibitory effect of TGFβ1 on *Tbx21* expression is mediated through SHP-1 [25], a negative regulator of TCR signaling and Treg cell generation. Our studies suggest that TGFβ1 inhibits Th1 differentiation through a Ca^{2+}/calcineurin pathway, but it is less clear whether that is the pathway used to inhibit Th2 differentiation [13]. It is also possible that, depending on the cytokine environment, TGFβ1 induces either Th3 cells or Th17 cells, thereby causing an indirect suppression of Th1 and Th2 differentiation.

E3 ubiquitin ligase CBLB is a negative regulator of T cell activation, and a deficiency in CBLB leads to proliferation of T cells independently of CD28-mediated costimulation. CBLB-deficient mice are also more prone to the induction of autoimmune diseases [26]. Interestingly, T cell responsiveness is not inhibitable by TGFβ1 in the absence of CBLB, suggesting that TGFβ1 may exert some of its immunoregulatory functions through CBLB in T cells. *In vitro* induction of *Foxp3* in activated T cells by TGFβ1 is largely dependent on *Cblb* expression [18]. However, CBLB-deficient mice live longer and do not exhibit any T cell activation as compared to *Tgfb1*$^{-/-}$ mice, suggesting that TGFβ1 may inhibit T cell activation through other molecules [26]. In CBLB-deficient T cells activation of VAV1 (guanine nucleotide exchange factor) was increased and SMAD2 phosphorylation was decreased, suggesting that the TGFβ1-mediated regulatory effect on T cells is partly mediated by CBLB [26]. VAV1 is an adapter protein activated by TCR stimulation, which is essential for some TCR signaling events such as Ca^{2+} flux, cytokine production and proliferation. VAV1-deficient T cells are hyporesponsive to mitogenic stimulation, and genetic ablation of *Cblb* restores proliferative competence of *Vav1*$^{-/-}$ T cells in the absence of Ca^{2+} flux. It has been shown that CBLB is degraded upon stimulation of T cells with anti-CD3 plus anti-CD28 antibodies, whereas anti-CTLA-4 antibody treatment induces *Cblb* expression, suggesting that costimulatory signals through CD28 and CTLA-4 have opposing effects on *Cblb*. These data imply that CBLB inhibits VAV1 activity and that a stronger stimulation through CD28 overrides CBLB activity by degrading it [27].

Smad3$^{-/-}$ mice do not develop autoimmune disease, but exhibit a mild activation of T cells and mucosal hyperplasia leading to colon cancer [28]. Adult mice with *Smad4* conditional knockout (CKO) in T cells exhibit a decrease in CD25 expression on T cells, which produce the increased amounts of proinflammatory cytokines IL-6, IL-4, IL-5 and IL-13 (Th2 cytokines) that cause the inflammation and colon cancer [29]. Akt/PKB also binds to SMAD3 and prevents its phosphorylation by TGFβ receptor kinases [30]. These data suggest that TGFβ1 utilizes distinct pathways in the same cell type to exert its multiple effects, and that SMAD-independent signaling mechanisms play an important role in T cells, B cells and other cell types [30a].

Phenotypes of TGFβ1-deficient mice

Tgfb1$^{-/-}$ mice that survive to birth die within a month after birth from a multifocal autoimmune disease caused by self-reactive T cells [1, 31]. Phenotypes of *Tgfb1*$^{-/-}$ mice on various gene knockout backgrounds are summarized in a recent review [30a]. Elimination of T cells or IL-6 prolongs survival of *Tgfb1*$^{-/-}$ mice by 2–3 months, suggesting that TGFβ1-deficient T cells cause inflammation through production of IL-6. *Tgfb1*$^{-/-}$ *Il6*$^{-/-}$ mice live longer than *Tgfb1*$^{-/-}$ *Ifng*$^{-/-}$ mice ([32]; T. Doetschman, unpublished observation). In *Tgfb1*$^{-/-}$ *Ifng*$^{-/-}$ mice, inflammation is eliminated only in the liver but not in the heart and lungs, whereas inflammation is either eliminated or reduced in the majority of tissues in *Tgfb1*$^{-/-}$ *Il6*$^{-/-}$ mice. This suggests that pathogenic Th17 cells may be induced by IL-6 in the absence of IFN-γ and cause disease in *Tgfb1*$^{-/-}$ *Ifng*$^{-/-}$ mice, leading to early lethality. Consistent with the recent observations that TGFβ1 and IL-6 are essential for Th17 cell differentiation, Th17 cells are absent in *Tgfb1*$^{-/-}$ mice, suggesting that *Tgfb1*$^{-/-}$ T cells cause autoimmune disease without differentiating into Th17 cells [8, 33]. Since *Tgfb1*$^{-/-}$ T cells produce IFN-γ but not IL-17, it is possible that IFN-γ produced by Th1 cells inhibits Th17 cell development in *Tgfb1*$^{-/-}$ mice [14, 32, 33]. Since *Tgfb1*$^{-/-}$ mice produce increased amounts of IFN-γ, it is possible that Th1 cells cause disease in *Tgfb1*$^{-/-}$ mice. Since we have observed that in *Tgfb1*$^{-/-}$ mice CD8 T cells also cause inflammation without CD4 helper T cells, it is unclear whether the inflammation is caused by Th17 or CD8 T cells in *Tgfb1*$^{-/-}$ *Ifng*$^{-/-}$ mice [30a, 32, 34].

Treatment of *Tgfb1*$^{-/-}$ mice from day 2 after birth with CTLA-4-Ig, which binds to B7.1 with high affinity and blocks costimulation through both CD28 and CTLA-4 [35], does not reverse the severe autoimmune disease (T. Doetschman, S. J. Engle, D. A. Rowley, J. A. Bluestone, unpublished observation) [36]. These data suggest that TGFβ1-deficient T cells do not require costimulatory signals through CD28 for their activation and autoimmune response [1]. Treatment of *Tgfb1*$^{-/-}$ newborn pups with anti-LFA-1 antibody every other day extends the survival of these pups as long as they are treated, but they succumb to inflammation within 1 week after treatment termination [37]. We have shown that LFA-1 (a costimulatory adhesion molecule involved in T cell migration and extravasation) is up-regulated on *Tgfb1*$^{-/-}$ T cells, so the antibody treatment may block LFA-1 interaction with its ligands [14, 37].

Our studies suggest that both $CD4^+$ and $CD8^+$ T cells are capable of causing disease in *Tgfb1*$^{-/-}$ mice, and that elimination of lymphocytes (RAG deficiency) or all T cells, but not B cells alone, is sufficient to eliminate inflammation [34]. The inflammatory disease mediated by T cells in *Tgfb1*$^{-/-}$ mice is not due to their response to enteric flora, because the inflammation is not reduced by elimination of enteric flora in germ-free *Tgfb1*$^{-/-}$ mice [38]. These data suggest that T cells in *Tgfb1*$^{-/-}$ mice may respond to self antigens, thus leading to activation and autoimmunity. These results clearly indicate an essential role for TGFβ1 in T cell regulation and autoimmune disease, but they do not elucidate the mechanism(s) of TGFβ1 function in T cells.

We have found that TGFβ1 prevents autoimmune disease by elevating the threshold level of T cell activation through Ca^{2+}-calcineurin signaling [13]. *Tgfb1*$^{-/-}$ thymocytes from mice with no inflammation have elevated $[Ca^{2+}]_i$ levels and exhibit an activated phenotype after suboptimal stimulation. Addition of TGFβ1 to $CD4^+$ T cells during activation with anti-$CD3^+$ and anti-CD28 prevents Ca^{2+} influx, NFATc activation and nuclear translocation of NFATc, which are important for cytokine production and naïve T cell proliferation [15]. Increased activation-induced cell death (AICD) of *Tgfb1*$^{-/-}$ T cells is due to a T cell-intrinsic deficiency of TGFβ1, because exogenous TGFβ1 does not prevent AICD, and induction of the survival factor *Bcl2l1* upon activation of T cells is greatly reduced in *Tgfb1*$^{-/-}$ T cells [20]. This protective effect was also suggested to be SMAD3 independent as SMAD3-deficient T cells also respond to TGFβ1 [39]. Hyperresponsiveness of TGFβ1-deficient T cells seems to be SMAD4 independent since SMAD4 deletion in T cells does not lead to autoimmune disease [29].

TGFβ in Treg cell function

TGFβ1 plays an important role in $CD4^+CD25^+$ Treg cell function. The TGFβ signaling mechanism is more complex than we think because multiple isoforms of TGFβ bind to the same receptor but signal through multiple co-receptors. Multiple subsets of T cells exhibit regulatory function *in vitro* and *in vivo*. $CD4^+CD25^+$ and $CD4^+CD25^-$ T cells exhibit suppressor function on naïve T cells, and their suppressor function is correlated with *Foxp3* expression. The molecular mechanism by which Treg cells inhibit T cell response is now becoming clear with recent observations made in *Tgfbr2* CKO mice [4, 5, 40].

Recently, it has been found that those T cells with a TGFβR2 deficiency, in mice with *Lck/Cre*-mediated CKO of *Tgfbr2* in T cells, are resistant to Treg cell suppression and have a T cell phenotype similar to *Tgfb1*$^{-/-}$ mice [4, 5]. Blockade of TGFβ1 using neutralizing antibody also abolishes suppressor function of TGFβ1-deficient Treg cells *in vivo* in adoptive transfer studies, suggesting that paracrine TGFβ1 produced within recipient mice may mediate the suppressor function of TGFβ1-deficient Treg cells [41]. Since *Tgfb1*$^{-/-}$ mice die around weaning age from multi-organ autoimmune inflammatory diseases, it is very unlikely that TGFβ1-independent Treg cells have the ability to induce dominant tolerance *in vivo* [1, 31].

A decrease in functional Treg cells results in autoimmune diseases such as type 1 diabetes and inflammatory bowel disease [42]. *TGFB1* overexpressed in Th3 cells rescues IL-2-deficient mice from autoimmune disease and also protects mice from EAE [43]. TGFβ1 induces *Foxp3* in T cells, which in turn inhibits *Smad7* (inhibitory

SMAD) in Treg cells in a positive autoregulatory manner [9]. This suggests that the TGFβ1 produced by Th3 cells may induce the conversion of Th cells to Treg cells in a cell-cell contact-dependent manner (infectious tolerance) [44].

Our recent studies suggest that *Foxp3* expression is not directly regulated by TGFβ1 because generation and maintenance of $CD4^+CD25^+$ Treg cells ($FOXP3^+$) are actually increased in transgenic TCR-expressing (DO11.10) *Tgfb1*$^{-/-}$ mice. CTLA-4 is required for expression of *Foxp3* in activated T cells in the presence of TGFβ1, although it is not required for *Foxp3* expression in natural Treg cells that develop in the thymus [45]. However, expression of FOXP3 alone in Treg cells is not sufficient for their function since Treg cells in *Tgfb1*$^{-/-}$ mice are unable to prevent activation of $CD4^+CD25^-$ T cells, suggesting that deficiency of TGFβ1 probably makes the Treg cells defective in controlling the activation of T cells [45a].

Mouse models in which TGFβ signaling is disrupted develop cell-autonomous autoimmune disease [30a], suggesting an important physiological role for TGFβ1-dependent Treg cells. In NOD mice the proportion of TGFβ1-producing Treg cells decreases with age and correlates with disease onset and progression [46]. TGFβ signaling-deficient T cells are not inhibited by Treg cells *in vivo*, suggesting that TGFβ-dependent Treg cells are major players in tolerance induction *in vivo* [47]. Treg cell numbers are reduced in mice either lacking TGFβR2 or expressing a *dn-Tgfbr2* in T cells [4, 5]. These data suggest that autocrine TGFβ1 in Treg cells is important for their maintenance, expansion and suppressor function. This is also supported by the finding that *Tgfb1*$^{-/-}$, *Tgfbr2* CKO, and *dn-Tgfbr2* transgenic mice all exhibit similar autoimmune phenotypes, with the delay of onset in the latter being due to transgene leakiness [30a]. Together, these *in vivo* and *in vitro* studies suggest that TGFβ1-deficient Treg cells are not effective in a TGFβ-deficient environment [11, 41, 45a]. Further studies are required to understand the molecular mechanisms that control the expression of TGFβ1 in Treg cells during infection and inflammatory diseases.

Treg cells produce TGFβ1, and blocking TGFβ1 with neutralizing antibody prevents the inhibitory action of Treg cells on T cells [48]. Cross-linking of CTLA-4 induces TGFβ1 production by $CD4^+$ T cells, and antibodies to CTLA-4 block Treg cell function *in vivo* [49]. These observations suggests that TGFβ1 production is both tightly regulated and important for $CD4^+CD25^+CTLA\text{-}4^+$ Treg cell function *in vivo* [48, 50]. It has also been shown that CTLA-4 plays an important role in the suppressor function of Treg cells independently of TGFβ1, although CTLA-4-deficient Treg cells utilize TGFβ1 as a compensatory mechanism [51].

The role of activation signal strength in T cell tolerance

Stimulation of T cells with ionomycin alone leads to anergy in the absence of PKC activation. Stimulation of T cells with PMA leads to activation of NF-κB and AP-1

(FOS/JUN), which prevents anergy induction by ionomycin [52]. FOXP3$^+$ Treg cells are anergic to stimulation by anti-CD3 and anti-CD28. TGFβ1 induces the conversion of naïve T cells to Treg cells expressing *Foxp3* in the presence of anti-CD3 and anti-CD28 [9]. These data suggest that activation signal strength instructs T cells either to become effector T cells or to become anergic (a state of unresponsiveness), and that TGFβ1 modulates TCR signal strength through its positive effects on CBLB and CTLA-4 production and its negative effects on calcineurin [13, 15, 53]. Blockade of costimulatory signals through CD4 or CD40L induce peripheral tolerance. CD4$^+$CD25$^+$ Treg cells can be generated from CD4$^+$CD25$^-$ T cells *in vivo* after treatment with non-mitogenic anti-CD3 F(ab')$_2$ fragments, suggesting that tolerance and immune response to antigens are regulated by signal strength [44, 54]. Similarly, another study by Ochi et al. [55] using oral therapy of anti-CD3, which induces LAP$^+$CD4$^+$ T cells confirms that TGFβ1-dependent Treg cell generation is important for preventing autoimmune disease.

DC in antigen presentation and T cell tolerance

Costimulatory molecules are also important for the generation of natural Treg cells. Elimination of these molecules by gene ablation results in decreased numbers of Treg cells. Bluestone et al. [56] demonstrated that Treg cells directly interact with antigen-presenting cells (APC) such as DC *in vivo* and prevent further arrest of effector T cells on DC. CTLA-4, which is constitutively expressed on Treg cells, binds to B7.1 (CD80) on APC, and down-modulates B7 molecules on DC *in vitro*; and these conditioned DC (DC that have encountered Treg cells) induce poor T cell proliferation [57]. Conditioned DC might produce the immunosuppressive cytokine TGFβ1, whereas those DC that encounter naïve T cells might produce the pro-inflammatory cytokine IL-12 and induce a Th1 response (Fig. 1). Consistently, CTLA-4 blockade results in decreased production of TGFβ1 and indoleamine 2,3-dioxigenase (IDO) and increased T cell response to viral infection [58]. Naïve T cell interaction with conditioned DC in a TGFβ1-rich environment could be tolerized rather than activated. Treg cell-produced TGFβ1 can act on both DC and naïve T cells and down-regulate their responses. TGFβ1 prevents maturation of DC by inhibiting costimulatory molecule expression [59].

DC were shown to express IDO upon interaction with CTLA-4, which is constitutively expressed on Treg cells [60]. IDO catabolizes tryptophan and leads to immunosuppression *via* T cell anergy [61]. Treg cells induce reverse signaling in DC upon their engagement with DC through CTLA-4 and its ligand CD80 on DC. Since Treg cells induce TGFβ1 in APC, it is possible there is signaling cross-talk in DC that leads to TGFβ1 production. CTLA-4-Ig treatment causes T cell anergy, which is dependent on IDO expression because IDO-deficient DC are not capable of inducing T cell tolerance [62]. However, it is unclear whether TGFβ1 production is also impaired in IDO-deficient DC. Since CTLA-4-Ig-treated *Tgfb1*$^{-/-}$ mice do not live longer than untreated *Tgfb1*$^{-/-}$ mice, and since they show no reduction in severity of inflammation as compared to that in control *Tgfb1*$^{-/-}$ mice ([36, 63]; and T. Doetschman; unpublished observation), we speculate that CTLA-4-Ig treatment induces tolerance through production of TGFβ1.

Summary

Several recent studies on Treg cells and the role of TGFβ1 in tolerance induction have led to the conclusion that TGFβ1 is central to self tolerance. TGFβ1 induces tolerance by raising the activation threshold levels of T cells, which prevents T cell activation by self-antigen presentation. A strong costimulatory environment overrides the suppressor function of TGFβ1 and allows T cells to respond to invading pathogens. Treg cells play an important role in maintaining lymphocyte homeostasis, and perturbations in Treg cell generation or function lead to autoreactive T cell expansion. The majority of Treg cells depend on TGFβ1 for their suppressor function, although they do not need to produce it. However, TGFβ signaling is essential for T cells to respond to Treg cell function. Current studies focusing on the role of TGFβ1 in T cell subsets using CKO mice will help to clarify the complex roles played by TGFβ1 in these processes. Present approaches in restoring the Treg cell pool using antibody therapy or gene transfer therapy will lead to better treatment options for autoimmune diseases.

Acknowledgements
We apologize to those investigators whose works have not been cited in this review owing to space restrictions. We thank Dr. Greg Boivin, University of Cincinnati, for his contributions to the IL-6 unpublished studies mentioned in the review, and we thank Dr. George Babcock for consultation and use of his FACS equipment. We acknowledge the support of NIH grants AI067903 and CA084291 to T.D.

References

1 Bommireddy R, Doetschman T (2004) TGF-beta, T-cell tolerance and anti-CD3 therapy. *Trends Mol Med* 10: 3–9
2 Li MO, Wan YY, Sanjabi S, Robertson AK, Flavell RA (2006) Transforming growth factor-beta regulation of immune responses. *Annu Rev Immunol* 24: 99–146
3 Schmidt-Weber CB, Blaser K (2004) Regulation and role of transforming growth fac-

tor-beta in immune tolerance induction and inflammation. *Curr Opin Immunol* 16: 709–716

4 Marie JC, Liggitt D, Rudensky AY (2006) Cellular mechanisms of fatal early-onset autoimmunity in mice with the T cell-specific targeting of transforming growth factor-beta receptor. *Immunity* 25: 441–454

5 Li MO, Sanjabi S, Flavell RA (2006) Transforming growth factor-beta controls development, homeostasis, and tolerance of T cells by regulatory T cell-dependent and -independent mechanisms. *Immunity* 25: 455–471

6 Bettelli E, Carrier Y, Gao W, Korn T, Strom TB, Oukka M, Weiner HL, Kuchroo VK (2006) Reciprocal developmental pathways for the generation of pathogenic effector TH17 and regulatory T cells. *Nature* 441: 235–238

7 Chen W, Wahl SM (2002) TGF-beta: receptors, signaling pathways and autoimmunity. *Curr Dir Autoimmun* 5:62–91

8 Weaver CT, Harrington LE, Mangan PR, Gavrieli M, Murphy KM (2006) Th17: an effector CD4 T cell lineage with regulatory T cell ties. *Immunity* 24: 677–688

9 Fantini MC, Becker C, Monteleone G, Pallone F, Galle PR, Neurath MF (2004) Cutting edge: TGF-beta induces a regulatory phenotype in $CD4^+CD25^-$ T cells through Foxp3 induction and down-regulation of Smad7. *J Immunol* 172: 5149–5153

10 Chen W, Jin W, Hardegen N, Lei KJ, Li L, Marinos N, McGrady G, Wahl SM (2003) Conversion of peripheral $CD4^+CD25^-$ naive T cells to $CD4^+CD25^+$ regulatory T cells by TGF-beta induction of transcription factor Foxp3. *J Exp Med* 198: 1875–1886

11 Marie JC, Letterio JJ, Gavin M, Rudensky AY (2005) TGF-beta1 maintains suppressor function and Foxp3 expression in $CD4^+CD25^+$ regulatory T cells. *J Exp Med* 201: 1061–1067

12 Zheng SG, Wang JH, Stohl W, Kim KS, Gray JD, Horwitz DA (2006) TGF-beta requires CTLA-4 early after T cell activation to induce FoxP3 and generate adaptive $CD4^+CD25^+$ regulatory cells. *J Immunol* 176: 3321–3329

13 Bommireddy R, Ormsby I, Yin M, Boivin GP, Babcock GF, Doetschman T (2003) TGF-beta1 inhibits Ca^{2+}-calcineurin-mediated activation in thymocytes. *J Immunol* 170: 3645–3652

14 Bommireddy R, Saxena V, Ormsby I, Yin M, Boivin GP, Babcock GF, Singh RR, Doetschman T (2003) TGF-beta1 regulates lymphocyte homeostasis by preventing activation and subsequent apoptosis of peripheral lymphocytes. *J Immunol* 170: 4612–4622

15 Chen CH, Seguin-Devaux C, Burke NA, Oriss TB, Watkins SC, Clipstone N, Ray A (2003) Transforming growth factor beta blocks Tec kinase phosphorylation, Ca^{2+} influx, and NFATc translocation causing inhibition of T cell differentiation. *J Exp Med* 197: 1689–1699

16 Wakefield LM, Roberts AB (2002) TGF-beta signaling: positive and negative effects on tumorigenesis. *Curr Opin Genet Dev* 12: 22–29

17 Wolfraim LA, Walz TM, James Z, Fernandez T, Letterio JJ (2004) p21Cip1 and

p27Kip1 act in synergy to alter the sensitivity of naive T cells to TGF-beta-mediated G1 arrest through modulation of IL-2 responsiveness. *J Immunol* 173: 3093–3102

18 Chipuk JE, Bhat M, Hsing AY, Ma J, Danielpour D (2001) Bcl-xL blocks transforming growth factor-beta 1-induced apoptosis by inhibiting cytochrome c release and not by directly antagonizing Apaf-1-dependent caspase activation in prostate epithelial cells. *J Biol Chem* 276: 26614–26621

19 Sung JL, Lin JT, Gorham JD (2003) CD28 co-stimulation regulates the effect of transforming growth factor-beta1 on the proliferation of naive CD4⁺ T cells. *Int Immunopharmacol* 3: 233–245

20 Chen W, Jin W, Tian H, Sicurello P, Frank M, Orenstein JM, Wahl SM (2001) Requirement for transforming growth factor beta1 in controlling T cell apoptosis. *J Exp Med* 194: 439–453

21 Khanna AK, Hosenpud JD (1999) Cyclosporine induces the expression of the cyclin inhibitor p21. *Transplantation* 67: 1262–1268

22 Ivanov II, McKenzie BS, Zhou L, Tadokoro CE, Lepelley A, Lafaille JJ, Cua DJ, Littman DR (2006) The orphan nuclear receptor RORgammat directs the differentiation program of proinflammatory IL-17(+) T helper cells. *Cell* 126: 1121–1133

23 Lin JT, Martin SL, Xia L, Gorham JD (2005) TGF-beta 1 uses distinct mechanisms to inhibit IFN-gamma expression in CD4⁺ T cells at priming and at recall: differential involvement of Stat4 and T-bet. *J Immunol* 174: 5950–5958

24 Gorelik L, Fields PE, Flavell RA (2000) Cutting edge: TGF-beta inhibits Th type 2 development through inhibition of GATA-3 expression. *J Immunol* 165: 4773–4777

25 Park IK, Shultz LD, Letterio JJ, Gorham JD (2005) TGF-beta1 inhibits T-bet induction by IFN-gamma in murine CD4⁺ T cells through the protein tyrosine phosphatase Src homology region 2 domain-containing phosphatase-1. *J Immunol* 175: 5666–5674

26 Chiang YJ, Kole HK, Brown K, Naramura M, Fukuhara S, Hu RJ, Jang IK, Gutkind JS, Shevach E, Gu H (2000) Cbl-b regulates the CD28 dependence of T-cell activation. *Nature* 403: 216–220

27 Krawczyk C, Bachmaier K, Sasaki T, Jones RG, Snapper SB, Bouchard D, Kozieradzki I, Ohashi PS, Alt FW, Penninger JM (2000) Cbl-b is a negative regulator of receptor clustering and raft aggregation in T cells. *Immunity* 13: 463–473

28 Zhu Y, Richardson JA, Parada LF, Graff JM (1998) Smad3 mutant mice develop metastatic colorectal cancer. *Cell* 94: 703–714

29 Kim BG, Li C, Qiao W, Mamura M, Kasperczak B, Anver M, Wolfraim L, Hong S, Mushinski E, Potter M, Kim SJ, Fu XY, Deng C, Letterio JJ (2006) Smad4 signalling in T cells is required for suppression of gastrointestinal cancer. *Nature* 441: 1015–1019

30 Conery AR, Cao Y, Thompson EA, Townsend CM Jr, Ko TC, Luo K (2004) Akt interacts directly with Smad3 to regulate the sensitivity to TGF-beta induced apoptosis. *Nat Cell Biol* 6: 366–372

30a Bommireddy R, Doetschman T (2007) TGFβ1 and Treg cells: alliance for tolerance. *Trends Mol Med* 13; *in press*

31 Bommireddy R, Pathak LJ, Martin J, Ormsby I, Engle SJ, Boivin GP, Babcock GF,

Eriksson AU, Singh RR, Doetschman T (2006) Self-antigen recognition by TGFbeta1-deficient T cells causes their activation and systemic inflammation. *Lab Invest* 86: 1008–1019

32 Gorham JD, Lin JT, Sung JL, Rudner LA, French MA (2001) Genetic regulation of autoimmune disease: BALB/c background TGF-beta 1-deficient mice develop necroinflammatory IFN-gamma-dependent hepatitis. *J Immunol* 166: 6413–6422

33 Mangan PR, Harrington LE, O'Quinn DB, Helms WS, Bullard DC, Elson CO, Hatton RD, Wahl SM, Schoeb TR, Weaver CT (2006) Transforming growth factor-beta induces development of the T(H)17 lineage. *Nature* 441: 231–234

34 Bommireddy R, Engle SJ, Ormsby I, Boivin GP, Babcock GF, Doetschman T (2004) Elimination of both CD4(+) and CD8(+) T cells but not B cells eliminates inflammation and prolongs the survival of TGFbeta1-deficient mice. *Cell Immunol* 232: 96–104

35 Manzotti CN, Liu MK, Burke F, Dussably L, Zheng Y, Sansom DM (2006) Integration of CD28 and CTLA-4 function results in differential responses of T cells to CD80 and CD86. *Eur J Immunol* 36: 1413–1422

36 Mamura M, Lee W, Sullivan TJ, Felici A, Sowers AL, Allison JP, Letterio JJ (2004) CD28 disruption exacerbates inflammation in TGFbeta1$^{-/-}$ mice: *in vivo* suppression by CD4$^+$CD25$^+$ regulatory T cells independent of autocrine TGF-beta1. *Blood* 103: 4594–4601

37 Diebold RJ, Eis MJ, Yin M, Ormsby I, Boivin GP, Darrow BJ, Saffitz JE, Doetschman T (1995) Early-onset multifocal inflammation in the transforming growth factor beta 1-null mouse is lymphocyte mediated. *Proc Natl Acad Sci USA* 92: 12215–12219

38 Boivin GP, Ormsby I, Jones-Carson J, O'Toole BA, Doetschman T (1997) Germ-free and barrier-raised TGF beta 1-deficient mice have similar inflammatory lesions. *Transgenic Res* 6: 197–202

39 McKarns SC, Schwartz RH (2005) Distinct effects of TGF-beta 1 on CD4$^+$ and CD8$^+$ T cell survival, division, and IL-2 production: a role for T cell intrinsic Smad3. *J Immunol* 174: 2071–2083

40 Miyara M, Sakaguchi S (2007) Natural regulatory T cells: mechanisms of suppression. *Trends Mol Med* 13: 108–116

41 Fahlen L, Read S, Gorelik L, Hurst SD, Coffman RL, Flavell RA, Powrie F (2005) T cells that cannot respond to TGF-beta escape control by CD4(+)CD25(+) regulatory T cells. *J Exp Med* 201: 737–746

42 Gregg RK, Jain R, Schoenleber SJ, Divekar R, Bell JJ, Lee HH, Yu P, Zaghouani H (2004) A sudden decline in active membrane-bound TGF-beta impairs both T regulatory cell function and protection against autoimmune diabetes. *J Immunol* 173: 7308–7316

43 Carrier Y, Yuan J, Kuchroo VK, Weiner HL (2007) Th3 cells in peripheral tolerance. I. Induction of Foxp3-positive regulatory T cells by Th3 cells derived from TGF-beta T cell-transgenic mice. *J Immunol* 178: 179–185

44 Graca L, Chen TC, Le Moine A, Cobbold SP, Howie D, Waldmann H (2005) Dominant tolerance: activation thresholds for peripheral generation of regulatory T cells. *Trends Immunol* 26: 130–135

45 Zheng SG, Wang JH, Gray JD, Soucier H, Horwitz DA (2004) Natural and induced CD4^{+}CD25^{+} cells educate CD4^{+}CD25^{-} cells to develop suppressive activity: the role of IL-2, TGF-beta, and IL-10. *J Immunol* 172: 5213–5221
45a Bommireddy R, Babcock GF, Singh RR, Doetschman T. TGFβ1 is not essential for natural Treg cell generation and maintenance. *Clin Immunol; in revision*
46 Pop SM, Wong CP, Culton DA, Clarke SH, Tisch R (2005) Single cell analysis shows decreasing FoxP3 and TGF-beta1 coexpressing CD4^{+}CD25^{+} regulatory T cells during autoimmune diabetes. *J Exp Med* 201: 1333–1346
47 Green EA, Gorelik L, McGregor CM, Tran EH, Flavell RA (2003) CD4^{+}CD25^{+} T regulatory cells control anti-islet CD8^{+} T cells through TGF-beta-TGF-beta receptor interactions in type 1 diabetes. *Proc Natl Acad Sci USA* 100: 10878–10883
48 Nakamura K, Kitani A, Strober W (2001) Cell contact-dependent immunosuppression by CD4(+)CD25(+) regulatory T cells is mediated by cell surface-bound transforming growth factor beta. *J Exp Med* 194: 629–644
49 Chen W, Jin W, Wahl SM (1998) Engagement of cytotoxic T lymphocyte-associated antigen 4 (CTLA-4) induces transforming growth factor beta (TGF-beta) production by murine CD4(+) T cells. *J Exp Med* 188: 1849–1857
50 Belghith M, Bluestone JA, Barriot S, Megret J, Bach JF, Chatenoud L (2003) TGF-beta-dependent mechanisms mediate restoration of self-tolerance induced by antibodies to CD3 in overt autoimmune diabetes. *Nat Med* 9: 1202–1208
51 Tang Q, Boden EK, Henriksen KJ, Bour-Jordan H, Bi M, Bluestone JA (2004) Distinct roles of CTLA-4 and TGF-beta in CD4^{+}CD25^{+} regulatory T cell function. *Eur J Immunol* 34: 2996–3005
52 Macian F, Garcia-Cozar F, Im SH, Horton HF, Byrne MC, Rao A (2002) Transcriptional mechanisms underlying lymphocyte tolerance. *Cell* 109: 719–731
53 Wohlfert EA, Gorelik L, Mittler R, Flavell RA, Clark RB (2006) Cutting Edge: Deficiency in the E3 ubiquitin ligase Cbl-b results in a multifunctional defect in T cell TGF-beta sensitivity *in vitro* and *in vivo*. *J Immunol* 176: 1316–1320
54 Chatenoud L (2005) CD3-specific antibodies restore self-tolerance: mechanisms and clinical applications. *Curr Opin Immunol* 17: 632–637
55 Ochi H, Abraham M, Ishikawa H, Frenkel D, Yang K, Basso AS, Wu H, Chen ML, Gandhi R, Miller A, Maron R, Weiner HL (2006) Oral CD3-specific antibody suppresses autoimmune encephalomyelitis by inducing CD4(+)CD25(−)LAP(+) T cells. *Nat Med* 12: 627–635
56 Tang Q, Adams JY, Tooley AJ, Bi M, Fife BT, Serra P, Santamaria P, Locksley RM, Krummel MF, Bluestone JA (2006) Visualizing regulatory T cell control of autoimmune responses in nonobese diabetic mice. *Nat Immunol* 7: 83–92
57 Oderup C, Cederbom L, Makowska A, Cilio CM, Ivars F (2006) Cytotoxic T lymphocyte antigen-4-dependent down-modulation of costimulatory molecules on dendritic cells in CD4^{+} CD25^{+} regulatory T-cell-mediated suppression. *Immunology* 118: 240–249
58 Hryniewicz A, Boasso A, Edghill-Smith Y, Vaccari M, Fuchs D, Venzon D, Nacsa J,

Betts MR, Tsai WP, Heraud JM et al (2006) CTLA-4 blockade decreases TGF-beta, IDO, and viral RNA expression in tissues of SIVmac251-infected macaques. *Blood* 108: 3834–3842

59 Lyakh LA, Sanford M, Chekol S, Young HA, Roberts AB (2005) TGF-beta and vitamin D3 utilize distinct pathways to suppress IL-12 production and modulate rapid differentiation of human monocytes into $CD83^{+}$ dendritic cells. *J Immunol* 174: 2061–2070

60 Grohmann U, Orabona C, Fallarino F, Vacca C, Calcinaro F, Falorni A, Candeloro P, Belladonna ML, Bianchi R, Fioretti MC, Puccetti P (2002) CTLA-4-Ig regulates tryptophan catabolism *in vivo*. *Nat Immunol* 3: 1097–1101

61 Munn DH, Sharma MD, Baban B, Harding HP, Zhang Y, Ron D, Mellor AL (2005) GCN2 kinase in T cells mediates proliferative arrest and anergy induction in response to indoleamine 2,3-dioxygenase. *Immunity* 22: 633–642

62 Mellor AL, Baban B, Chandler P, Marshall B, Jhaver K, Hansen A, Koni PA, Iwashima M, Munn DH (2003) Cutting edge: induced indoleamine 2,3 dioxygenase expression in dendritic cell subsets suppresses T cell clonal expansion. *J Immunol* 171: 1652–1655

63 Tivol EA, Boyd SD, McKeon S, Borriello F, Nickerson P, Strom TB, Sharpe AH (1997) CTLA4Ig prevents lymphoproliferation and fatal multiorgan tissue destruction in CTLA-4-deficient mice. *J Immunol* 158: 5091–5094

From mice to men: the challenges of developing tolerance-inducing biological drugs for the clinic

Wan-Fai Ng and John D. Isaacs

Musculoskeletal Research Group, Institute of Cellular Medicine, University of Newcastle, Newcastle upon Tyne, NE2 4HH, UK

The road to tolerance-inducing therapies

Before discussing the specific barriers to clinical therapeutic tolerance, it is useful to review the steps involved in biological drug development in general. The first is to identify an appropriate therapeutic target(s) based on current understanding of disease pathogenesis, which may be molecular or cellular in nature: an appropriately designed biologic that binds this target might reasonably be expected to influence the disease process. So-called "proof of concept" studies are then developed, typically in rodent models, to assess efficacy, safety, toxicity, immunogenicity, pharmacokinetics and metabolism of the newly developed biological drug. By their very nature, however, biological drugs that recognize animal targets generally will not recognize the equivalent human molecule. For instance, a monoclonal antibody (mAb) that binds murine CD3 will not recognize the human molecule due to inter-species differences in CD3 structure. Consequently, a different mAb must be developed for clinical studies, which will first be tested in *in vitro* and *ex vivo* studies using human cells or tissues, as well as in non-human primate models. Finally, the therapy will be evaluated in clinical trials.

The development of tumor necrosis factor (TNF) blockade for the treatment of rheumatoid arthritis (RA) is an excellent example of the rational development of a biological therapy. TNF was shown to be present in abundance in inflamed joints of patients with RA [1]. It was also observed that an anti-TNF mAb inhibited the production of pro-inflammatory cytokines by RA synovial cells *in vitro* [2]. These observations suggested that TNF played a central role in mediating joint inflammation. Anti-TNF mAbs were then tested in murine collagen-induced arthritis (a model for RA) and shown to be effective in reducing joint inflammation and damage [3]. An mAb suitable for use in humans was then developed, and its efficacy in RA subsequently confirmed in formal clinical trials [4, 5]. TNF blockade is now one of the most effective therapies available for RA.

In the following sections, we highlight the different challenges faced at each stage of development of tolerance-inducing biological therapies. These are summarized in Table 1.

The Immune Synapse as a Novel Target for Therapy, edited by Luis Graca

Table 1 - Key challenges in the development of biological tolerance-inducing therapies

Stage of Development	Key challenge(s)	Potential/existing barriers
Designing a novel therapy	Identify appropriate therapeutic target(s)	Pathogenesis of many autoimmune diseases is poorly understood Multiplicity of pathogenic pathways Disease heterogeneity between individuals
Rodent studies	Develop reliable predictive models of clinical outcome	Animal models are poor mimics of clinical disease, especially autoimmune conditions: a) They do not fully resemble human diseases b) Differences exist between immune systems of animals and humans c) Different lifespan – e.g., tolerance lasting >100 days in mice may not equate to life-time tolerance in humans d) Inbred animals vs. outbred humans e) "Germ-free" environment in animal models Different biological product must be developed for human studies, limiting extrapolation from rodent to man Pharmacokinetics, pharmacodynamics and immunogenicity may differ between animal and man Cannot precisely model Fc-FcγR interactions
Pre-clinical *in vitro* and *ex vivo* studies	Laboratory assays may not predict *in vivo* activity	Bench-top assays do not comprise all the cellular components that participate in the immune response *in vivo* Limited access to tissue from affected organs
Pre-clinical primate studies	Bioactivity of mAb on primate target may not be identical to that in human	Animal models are poor mimics of clinical disease, especially autoimmune conditions Ethical considerations limit number of animals that are treated Pharmacokinetics, pharmacodynamics and immunogenicity may differ Cannot precisely model Fc-FcγR interactions
Design of clinical trials	Development of appropriate biomarkers to guide therapy	Study cohort selection Effects of concomitant immunosuppressive treatment on tolerance induction unclear Ethical considerations How to assess tolerance a) Time course for tolerance induction unclear b) Lack of biomarker(s) c) Restoration of immunological tolerance may take several weeks during which inflammation is not suppressed

Challenges for achieving clinical therapeutic tolerance

Identifying an appropriate therapeutic target

The design of a specific, targeted biological therapy should ideally be based on a thorough understanding of the pathogenesis of the disease in question. Despite significant progress in understanding the mechanisms underlying transplant rejection, our knowledge of the pathogenesis of most human autoimmune diseases remains poor. Therefore, while we rely on animal models when designing biological therapies, these may be poorly informative. In contrast, the antigen(s) that initiate transplant rejection and the timing of the transplant procedure are clearly defined, and a more systematic approach to tolerance induction is possible here. Even in transplantation, however, the multiplicity of mechanisms mediating rejection and our incomplete understanding of "chronic rejection", render therapeutic target identification imperfect. Furthermore, there remains a significant gap in our knowledge of immunological tolerance as relevant to autoimmunity and transplantation, which adds to the difficulties of designing robust tolerance-inducing strategies.

Even when we seem to understand pathogenic mechanisms, targeted therapies may prove us wrong. For example, the success of TNF-α blockade in RA can be contrasted with the poorer efficacy of interleukin (IL)-1β blockade with anakinra, despite the fact that IL-1β appears to play a key role in joint inflammation in both human and rodent models of RA [6, 7]. Similarly, while anti-TNF-α mAbs are highly effective treatments for Crohn's disease [8], etanercept, a fusion protein of the p75 TNF-α receptor and human IgG_1 constant region (Fc), is ineffective [9]. More intriguingly, failure of one TNF-α antagonist in a patient with RA does not necessarily predict lack of efficacy of another [10–12]. The reasons for these discrepant findings are unclear but likely to be complex. Furthermore, agents that block cytokines are largely anti-inflammatory, and, while the above observations underscore the importance of appropriate target identification, this is even more difficult when dealing with a phenomenon such as immune tolerance where mechanisms are less well defined.

Preclinical testing

Animal models

Over the past decades, rodent models have served well to advance our knowledge of immunobiology. However, their role in the development of clinical therapeutic tolerance has been more limited. The reasons for their sometimes-poor predictive power can broadly be divided into factors that relate to the models themselves and to extrinsic factors.

Intrinsic limitations of animal models

Some of the most robust tolerogenic strategies in animal models have produced disappointing results in humans. For instance, mAbs targeting T cell surface antigens were consistently effective for tolerance induction and reversed both autoimmunity and transplant rejection in animal models [13, 14]. Their use in clinical studies was, however, generally disappointing [15]. These observations raise the possibility of inter-species functional differences in the targeted molecule, limiting relevant modeling. However, differences in cell surface antigen expression, and altered pharmacokinetics and pharmacodynamics of the biological agent are other possibilities.

Inbred mice or rats are used in experimental models, whereas patients have distinct genetic backgrounds. Furthermore, each patient has different co-morbidities and is influenced by various environmental factors. Such differences may account for inter-individual variations in treatment responses. For example, FcγRIIIa (CD16) polymorphisms influence clinical responses to infliximab therapy in Crohn's disease [16], and to rituximab treatment in systemic lupus erythematosus (SLE) [17]. FcγRIIIa polymorphisms may also predict responsiveness to infliximab in RA [18], although a more recent study did not identify such correlation [19]. Another example is the inter-individual variability in the repopulation of peripheral blood B cells following treatment with rituximab in patients with RA and SLE [20, 21]. The factors that are responsible for these inter-individual variations are not fully known.

In some animal models the induction of autoimmunity is strain specific. For instance, only certain mouse strains are susceptible to collagen-induced arthritis, raising questions regarding the generalizability of the data generated from such models to clinical disease. It is also noteworthy that, increasingly, many rodent models of autoimmunity and transplantation are designed specifically to address a unique aspect of the disease – such as genetic predisposition, environmental trigger or effector mechanism – by the use of transgenic, knockout/knockin technologies or induction with specific antigens. While these models allow more precise and detailed analysis of a particular aspect of the disease, interpreting data from such models requires great care and circumspection.

Due to the short lifespan of mice and pragmatism in experimental design, it is generally accepted that if an animal remains tolerant for greater than 100 days, then true immunological tolerance was achieved. However, this may not equate to lifetime tolerance in humans. Furthermore, histological examination is not generally used to confirm therapeutic tolerance induction, and it has now been shown that chronic rejection may be present in some situations. For instance, donor-specific transfusion in rats was shown to prolong kidney graft survival "indefinitely", although histology showed that most kidneys were undergoing chronic rejection [22]. Indeed, the use of donor-specific blood transfusion pre-transplantation in clinic failed to induce tolerance or confer any consistent beneficial effects on organ survival [23].

Most experimental animals are kept in a relatively clean, largely "germ-free" environment. Human beings, in contrast, are constantly exposed to a variety of pathogens, which may adversely affect tolerance induction in several ways. For example, infections may play a key role in initiating autoimmune responses and transplant rejection [24, 25]. Possible mechanisms incorporate cross-reactive lymphocytes that are both pathogens and host specific, and costimulatory signals that activate low-affinity autoantigen-specific T cells [25–27]. Furthermore, inflammatory cytokines may actively suppress regulatory mechanisms that maintain tolerance in the host. For instance, IL-6 has been shown to inhibit the suppressive function of $CD4^{+}CD25^{+}Foxp3^{+}$ regulatory T cells [28]. Recurrent exposure to pathogens also plays an important role in shaping the repertoire of T and B cells [29] and the memory cell pools are significantly larger in humans than in experimental mice [29, 30]. Compared with naïve T cells, memory T cells have a lower threshold of activation and different signal requirements [31, 32]. They can also undergo homeostatic proliferation, potentially rendering memory T cells less susceptible to tolerance-inducing strategies [33–35]. In this regard, Valujskikh et al. [34] showed that priming of allogeneic T cells prevented the effects of costimulation blockade on prolonged cardiac allograft survivals.

Extrinsic factors that limit the value of animal models

Biological therapies that are tested in animal models cannot usually be applied directly in humans. Consequently modification, or development of a new but related product, is necessary prior to clinical application. However, the modified therapy in humans may possess very different characteristics, resulting in unpredictable beneficial and adverse effects. Several factors may underlie the incongruous results between the animal and human experience. This is well illustrated by the Phase I clinical trial of the 'superagonist' anti-CD28 mAb, TGN1412. Treatment of mice with an 'equivalent' mAb led to selective expansion of $CD4^{+}CD25^{+}Foxp3^{+}$ regulatory T cells, and promoted tolerance induction [24, 36]. The clinical product was tested in non-human primates and found to be well tolerated [37]. When administered to healthy human volunteers, however, all six subjects developed multi-organ failure within 24 h. This was due to rapid release of pro-inflammatory cytokines through non-selective activation of T cells [38]. At least part of the explanation relates to the inability of TGN1412, an mAb of human IgG_4 isotype, to interact with cynomolgus monkey FcγRs, thereby limiting T cell activation in this species. This contrasts with its affinity for human FcγRI and the high-affinity variant of FcγRIIIa [39–41]. Historically, much less attention has been paid to the Fc of a mAb than its affinity for antigen, yet its affinity for FcγRs (and complement cascade components) is an important consideration. In addition to side effects, these interactions have an important bearing on target cell fate and mAb half-life (through interactions with the neonatal FcγR, FcRn) [42–44]. A particularly important consideration is that, even when targeting the same molecule, the biological effects of mAbs depend on their binding affinity, and the specific epitopes that they recognize as well as other properties. The clinical trials of anti-CD4 mAbs in RA provide examples that illustrate the diversity of biological consequences that may be elicited by different mAbs targeting the same molecule. Over the last 15 years, several different anti-CD4 mAbs were developed and studied in clinical trials, comprising murine, chimeric, primatized, humanized and fully human mAbs. These mAbs bound different epitopes on human CD4 and incorporated a variety of isotypes. Although the majority failed to provide significant clinical benefits in RA, their effects on recipient $CD4^+$ cells as well as adverse reactions varied considerably (reviewed in [15]). Some lysed $CD4^+$ cells, some caused modulation or CD4 shedding, whereas others simply coated the target cells. Because the critical characteristics of tolerogenic mAbs against murine CD4 were not defined, however, it was not possible to specifically design an mAb for clinical studies with a 'guaranteed' tolerogenic profile. Indeed, we still do not know whether such a profile exists, underscoring our uncertainties surrounding therapeutic tolerance induction.

Taken together, these observations suggest that animal models can be useful for demonstrating the feasibility of tolerogenic strategies but that they have limited value in helping to design products for clinical use. In recent years, much progress has been made in the development of 'humanized' mice. These express certain human antigens and may even, for example, possess elements of a human immune system (reviewed in [45]). These may provide better predictive models for drug discovery and development, but expression levels transgenes will differ from those in humans and, furthermore, they may not be appropriately regulated by, for example, pro-inflammatory cytokines.

Preclinical in vitro/in vivo *studies*

Preclinical *in vitro* and *ex vivo* assays (using human cells and tissues) provide important information about the therapeutic product but again do not reliably predict clinical responses *in vivo*. This is partly because no *in vitro* or *ex vivo* models fully reconstitute the environment in which immune responses take place. In addition, tissues from affected organs as well as lymphoid tissues are often inaccessible for use in pharmacodynamic modeling. As already suggested, *in vitro* correlates of tolerance induction are lacking, rendering it impossible to use bioassays to identify a tolerogenic 'profile' that will translate to clinical success. Furthermore, *in vitro* assays are often poor predictors of even relatively simple *in vivo* biological read-outs. For example, it has generally been extremely difficult to predict *in vivo* effector function from *in vitro* assays such as complement-dependent cytotoxicity (CDC) or antibody-dependent cell-mediated cytotoxicity (ADCC). Although both mechanisms are thought to underlie target cell depletion, *in vivo* cytotoxicity is often seen with mAbs that are impotent *in vitro*. For example, an IgG_4 anti-CD52 mAb caused long-lasting lymphopenia in humans despite poor *in vitro* bioactivity [46].

Immunogenicity

A key challenge to the development of biological drugs in general is immunogenicity and the elicitation of so-called anti-globulins (anti-Igs). The biological effects of anti-Igs depend on the epitope that is recognized by the mAb as well as features such as their isotype and affinity. For example, anti-Igs that interact with the binding site of a therapeutic mAb (anti-idiotypes) may neutralize its activity, while anti-Igs binding elsewhere may enhance its clearance. In either case, the efficacy of the biological therapy may be reduced or abolished (tachyphylaxis). In extreme cases sensitization may result in anaphylaxis, although this is fortunately rare and frequently anti-Igs exist without apparent clinical sequelae. In other cases, infusion reactions may occur in association with reduced efficacy, as has been observed in some patients treated with infliximab [47, 48]. Pre-medication with corticosteroid may be useful in preventing such reactions.

Many factors influence immunogenicity, including intrinsic structure and biological properties of the therapy, manufacturing processes, host-dependent factors and treatment protocols (Tab. 2). Strategies to minimize the development of anti-Igs include concomitant immunosuppressive treatment, high- *versus* low-dose therapy, continuous *versus* intermittent treatment, "humanization" of the antibody and use of "non cell-binding" variants of therapeutic antibodies [48–52]. In terms of tolerogenic therapy, immunogenicity should not interfere if the treatment is truly 'one-off'. Intermittent, infrequent treatment, however, is more likely to induce anti-Igs – as seen when infliximab is used intermittently in the treatment of Crohn's disease [50, 53, 54].

The immunogenicity of biological therapy is understandably difficult to predict from animal models and can only truly be assessed in the clinic. Even 'fully human' mAbs, derived from human Ig transgenic mice, will contain a sequence that is 'foreign' to most human recipients – idiotypes related to the binding site and allotypic variants in the V and C-regions [55]. The measurement and interpretation of anti-Igs also requires significant expertise, and the various assays available have not been standardized between laboratories.

In summary, animal models, *in vitro* and *ex vivo* studies can provide important information on the likely efficacy of biological therapies but several important aspects cannot reliably be predicted. Preclinical studies in primates generally allow *in vivo* studies on the ultimate clinical product but limitations remain: the target antigen may not be absolutely identical in its structure or expression pattern/level, and FcγR interactions will differ from the human situation. Furthermore, ethical issues (and cost) limit the number of animals that can be used for these studies. With this background, our lack of tolerance biomarkers is a particularly critical limitation for the development of tolerogenic therapies. Consequently, the only method to evaluate a tolerogenic (biological) therapy is through well-designed, carefully conducted and monitored clinical trials. However, a number of ethical and practical issues may prejudice trial design.

Table 2 - Factors that affect immunogenicity

(a) Intrinsic properties of the biological therapy	(i) The presence of "foreign" or neo-antigenic epitopes (ii) Precise chemical structure, e.g., glycosylation pattern (iii) Allotypes (iv) The target antigen
(b) Manufacturing process	(i) Excipients of the final product (ii) Quality control, e.g., impurities, endotoxin
(c) Recipient factors	(i) Immunocompetence of recipients (ii) Concomitant medical conditions (iii) Genetic variations, e.g., FcγR polymorphisms, MHC allotypes (iv) Disease under treatment
(d) Protocol-related	(i) Route of administration (ii) Frequency of administration (iii) Dose (iv) Concomitant treatment

Clinical trials

Clinical trial design

Over recent decades the treatment of autoimmunity and transplant rejection has improved significantly. Existing immunosuppressive drugs are not without side effects and also lack specificity but they are generally well tolerated and efficacious. Therefore, initial clinical trials of novel tolerogenic therapies focus on patients who have failed conventional treatments or, alternatively, on their use in conjunction with standard therapies. Both approaches may bias against the biological therapy. In the first instance, the patient's disease stage may play a significant part in determining response: thus, a patient rejecting their transplant in the face of optimal immunosuppression provides a tough scenario for a novel strategy. In the second instance, conventional immunosuppressants could interfere with tolerogenic mechanisms or signaling pathways, and their effect on immunological tolerance induction requires further investigation. For instance, ciclosporin inhibits apoptosis of activated allogeneic T cells as well as the development of T cell anergy and regulatory T cells in rodent models of transplantation [56–58]. In contrast, rapamycin appears to facilitate the expansion of $CD4^{+}CD25^{+}Foxp3^{+}$ regulatory T cells in mice and in humans [56, 59]. Thus, the choice of concomitant immunosuppressive regimen may have profound influence on the success of tolerance-inducing therapy. The use of background 'conventional' immunosuppression also requires a determination of the optimal time to withdraw these drugs. Without biomarkers of immune tolerance this is a difficult or impossible decision to make.

The stage of autoimmune disease may also influence the outcome of therapy. For example, there is an increasing body of evidence to suggest that early treatment of RA (in the first few months of disease) is more likely to result in sustained remission than the same treatment given later on [60]. It has also been suggested that, in later disease, tolerance induction may first require neutralization of the inflammatory milieu prior to administration of tolerogenic drug(s) [61]. In addition, many patients in the later stages of autoimmunity may have already suffered irreversible end-organ damage and it may be difficult to detect significant clinical benefit even if a tolerance-inducing strategy is "successful". This is best illustrated in insulin-dependent diabetes: a tolerogenic treatment will only be effective if the pancreas still contains a critical mass of functional, insulin-producing cells (β-cells).

The genetic constitution of an individual can also affect disease manifestations and responsiveness to treatment and trial design should, where possible, take this into consideration. For instance, FcγR polymorphisms may influence the risk:benefit ratio of particular mAb-based therapeutic strategies. Access to relevant analyses is currently limited but will become more widespread in the future.

Evaluation of clinical trials

In the absence of reliable biomarkers of tolerance, deciding whether a tolerant state has been achieved poses a considerable challenge. Our limited understanding of tolerance mechanisms means that we do not know what features to look for or even where to look. Should we expect to find clues in peripheral blood or should we focus on the target organ, or even the draining lymph nodes? Will tolerance induction be associated with entirely normal histology in the target organ or should we expect an active immunological process to maintain tolerance locally? A disadvantage of histological endpoints is the need for invasive procedure(s), suitable in certain circumstances but not others. Current searches for biomarkers relying on genomic and proteomic analyses of peripheral blood cells could be promising.

An important consideration when evaluating therapeutic tolerance is that this may occur over a period of time, perhaps weeks to months. Furthermore, the tolerogenic strategy may not directly impact on symptoms in the short term. Taking the scenario of RA, if a tolerogenic strategy takes 3 months to become fully effective but does not possess anti-inflammatory properties, it is unlikely to improve symptoms in the short term. In terms of a conventional 12-week trial design this drug will 'fail'. If it is anticipated that there may be a delay in onset of action then it would be unethical not to co-prescribe an anti-inflammatory agent or, alternatively, a conventional anti-rheumatic drug. Now, however, we face the same issues as highlighted in an earlier section: that the co-prescribed medications could themselves interfere with tolerance induction. This again emphasizes the absolute need for biomarkers of the tolerogenic process, that would allow us to recognize potential benefit even before this is evident clinically. Anecdotally, we have noted symptomatic improvements in RA patients several months and even years after apparently ineffective tolerogenic strategies. At that time these patients were usually taking standard RA therapies that had previously proven ineffective, suggesting that the experimental treatment had indeed modified the disease process, but in a way that had been unrecognized at the time of therapy (J. D. Isaacs, unpublished observations).

Useful strategies/Future directions

Having considered the barriers to the development of clinical therapeutic tolerance, it is fitting to turn to discussion of strategies that may help to overcome them.

Identifying biomarkers of immune tolerance

One key issue that has to be resolved is the identification of reliable tests or biomarkers that predict and/or detect immunological tolerance. Such markers will allow us to assess the efficacy of tolerogenic strategies more accurately and preferably during tolerance induction – before it is clinically evident. Importantly, they may help us to decide whether or not an apparently ineffective strategy could be effective if administered in a different manner: at higher dose, for longer, or without concomitant immunosuppression. When we use concomitant immunosuppression they should tell us when it is safe to withdraw it. Animal models may help us to identify biomarkers but technological advances associated with large-scale gene expression profiling, particularly advances in analytical bioinformatics, have massively increased the information available from clinical samples. A similar (and complementary) approach will be the use of proteomics, the large-scale analysis of protein expression. The ultimate manifestation of such advances in data handling is a systems biology approach, where all available data (encompassing genomics, proteomics, metabolomics, etc.) is integrated into predictive models. The ultimate success of such approaches will depend on choosing the appropriate cell population(s) for analysis but substantial investments are being made in this area [62].

Despite the potential drawbacks of *in vitro/ex vivo* assays alluded to earlier, these can ultimately provide valuable functional data, particularly in the context of clinical transplantation, when the identity of the 'auto'antigen(s) is/are known and immune responses can be followed prospectively. Furthermore, with the development of more sensitive assays, for instance in cytokine measurement, it is possible that clinical tolerance may be defined by a combination of such bioassays.

Studying naturally occurring "tolerant" patients

Perhaps the most convincing evidence for clinical therapeutic tolerance is the observation that some patients, albeit uncommon (with the exception of corneal transplant), appear to develop tolerance spontaneously following allotransplantation. This, of course, only becomes apparent upon intentional or covert withdrawal of immunosuppressive drugs. Most reported cases of allograft tolerance have occurred in relation to liver transplantation. Devlin and colleagues [63] reported a series of 18 liver transplant recipients whose immunosuppressive therapy was progressively withdrawn. At 3 years, 5 patients remained completely off immunosuppression. Successful drug withdrawal was associated with transplantation for non-immune-mediated liver disease, fewer HLA mismatches and fewer early rejection episodes. In a similar study, 18 out of 95 patients remained drug-free for 10 months to 4.8 years after withdrawal of immunosuppression [64]. Cases of spontaneous tolerance to other solid organs, however, are rare, and only became apparent when immunosuppression was withdrawn secondary to non-compliance, drug toxicity or malignancy [65, 66]. Careful immunological study of such tolerant patients may provide important clues as to how tolerance can be achieved in the clinic, as well as how patients should be monitored and selected for careful withdrawal of immunosuppression.

Biological therapy registries and collaborative research networks

Our experience with existing biological therapies has served to highlight their unpredictable efficacy, mechanisms of action and adverse effects [15]. Therefore, patients treated with these agents must be carefully monitored. Phase I clinical trials are unique studies for which no animal models or preclinical studies can provide sufficient guidance. Furthermore, unanticipated events (whether adverse or beneficial) are usually rare and may occur some time after treatment, outside the time course of conventional clinical trials. Establishment of national registries for patients receiving these biological therapies is an effective way to collect information on both beneficial and adverse effects of treatment and to perform post-marketing surveillance.

Among individual research groups there is considerable heterogeneity in the criteria used to define tolerance, animal models employed, and designs and outcome measures in clinical trials. Diversity in research methodologies can serve to broaden knowledge on the one hand, but can hinder comparison of different strategies on the other. The development of collaborative groups such as the Immune Tolerance Network [62] can facilitate design of multi-center clinical trials, promote research collaboration, standardization of bioassay techniques and knowledge exchange. Ultimately, it is the collective efforts of scientists and clinicians internationally that will lead us to the holy grail of immunological tolerance in clinic.

Clinical therapeutic tolerance *versus* true immunological tolerance

True immune tolerance implies stable allograft function or complete remission of autoimmunity in the absence of maintenance immunosuppression. To date, this goal has been elusive but Calne coined the term "prope (almost) tolerance". In the transplantation setting this refers to long-term graft survival and function that requires only a minimal, non-toxic dose of immunosuppression [67]. Control of autoimmunity using a similar level of immunosuppression may be viewed in a similar manner, or perhaps a requirement for brief, intermittent courses of immunomodulatory therapy administered annually or less frequently. Until clinical therapeutic tolerance can be predictably achieved these are useful outcomes for the patient because they limit adverse effects of immunosuppressive therapy and minimize hospital visits, thereby greatly enhancing quality of life. Such outcomes should, therefore, be considered when we design trials to test therapeutic tolerance induction.

Conclusion

Tremendous progress has been made in recent years in the development of tolerance-inducing therapy. However, several challenges still lie ahead before clinical therapeutic tolerance becomes a reality. Incomplete understanding of disease pathogenesis and mechanisms of tolerance induction continue to compromise the identification of appropriate therapeutic target(s). Furthermore, the limited predictive power of existing animal models and preclinical *in vitro*/*ex vivo* bioassays means that first-into-man studies carry considerable uncertainties and risks. Carefully conducted clinical trials remain the only way to evaluate the efficacy of a tolerance-inducing therapy, but these are difficult to design in the absence of reliable biomarkers to guide therapy. The study of spontaneously tolerant patients may provide valuable insights into the mechanisms of immunological tolerance and the establishment of national or international registries of biological therapy and collaborative research networks will maximize our chances of attaining the holy grail of clinical therapeutic tolerance.

References

1 Buchan G, Barrett K, Turner M, Chantry D, Maini RN, Feldmann M (1988) Interleukin-1 and tumour necrosis factor mRNA expression in rheumatoid arthritis: prolonged production of IL-1 alpha. *Clin Exp Immunol* 73: 449–455

2 Brennan FM, Chantry D, Jackson A, Maini R, Feldmann M (1989) Inhibitory effect of TNF alpha antibodies on synovial cell interleukin-1 production in rheumatoid arthritis. *Lancet* 2: 244–247

3 Williams RO, Feldmann M, Maini RN (1992) Anti-tumor necrosis factor ameliorates joint disease in murine collagen-induced arthritis. *Proc Natl Acad Sci USA* 89: 9784–9788

4 Lipsky PE, van der Heijde DM, St Clair EW, Furst DE, Breedveld FC, Kalden JR, Smolen JS, Weisman M, Emery P, Feldmann M et al (2000) Infliximab and methotrexate in the treatment of rheumatoid arthritis. Anti-Tumor Necrosis Factor Trial in Rheumatoid Arthritis with Concomitant Therapy Study Group. *N Engl J Med* 343: 1594–1602

5 Elliott MJ, Maini RN, Feldmann M, Kalden JR, Antoni C, Smolen JS, Leeb B, Breedveld FC, Macfarlane JD, Bijl H et al (1994) Randomised double-blind comparison of chimeric monoclonal antibody to tumour necrosis factor alpha (cA2) *versus* placebo in rheumatoid arthritis. *Lancet* 344: 1105–1110

6 Nixon R, Bansback N, Brennan A (2007) The efficacy of inhibiting tumour necrosis factor alpha and interleukin 1 in patients with rheumatoid arthritis: a meta-analysis and adjusted indirect comparisons. *Rheumatology (Oxford)* 46: 1140–1147

7 Cohen S, Hurd E, Cush J, Schiff M, Weinblatt M, Moreland L, Kramer J, Bear M, Rich W, McCabe D (2002) Treatment of rheumatoid arthritis with anakinra, a recombinant human interleukin-1 receptor antagonist, in combination with methotrexate: Results of a twenty-four-week, multicenter, randomised, double-blind, placebo-controlled trial. *Arthritis Rheum* 46: 614–624

8 Targan S, Hanauer S, van Deventer S, Mayer L, Present D, Braakman T, DeWoody K, Schaible T, Rutgeerts P (1997) A short-term study of chimeric monoclonal antibody cA2 to tumor necrosis factor for Crohn's disease. *N Engl J Med* 337: 1029–1036

9 Sandborn WJ, Hanauer SB, Katz S, Safdi M, Wolf DG, Baerg RD, Tremaine WJ, Johnson T, Diehl NN, Zinsmeister AR (2001) Etanercept for active Crohn's disease: a randomized, double-blind, placebo-controlled trial. *Gastroenterology* 121: 1088–1094

10 Nikas SN, Voulgari PV, Alamanos Y, Papadopoulos CG, Venetsanopoulou AI, Georgiadis AN, Drosos AA (2006) Efficacy and safety of switching from infliximab to adalimumab: a comparative controlled study. *Ann Rheum Dis* 65: 257–260

11 Hansen KE, Hildebrand JP, Genovese MC, Cush JJ, Patel S, Cooley DA, Cohen SB, Gangnon RE, Schiff MH (2004) The efficacy of switching from etanercept to infliximab in patients with rheumatoid arthritis. *J Rheumatol* 31: 1098–1102

12 van Vollenhoven R, Harju A, Brannemark S, Klareskog L (2003) Treatment with infliximab (Remicade) when etancercept (Enbrel) has failed or *vice versa*: data from the STURE registry showing that switching tumour necrosis factor alpha blockers can make sense. *Ann Rheum Dis* 62: 1195–1198

13 Waldmann H, Cobbold S (1998) How do monoclonal antibodies induce tolerance? A role for infectious tolerance? *Annu Rev Immunol* 16: 619–644

14 Cobbold S, Qin S, Leong L, Martin G, Waldmann H (1992) Reprogramming the immune system for peripheral tolerance with CD4 and CD8 monoclonal antibodies. *Immunol Rev* 129: 165–201

15 Strand V, Kimberly R, Isaacs JD (2007) Biologic therapies in rheumatology: lessons learned, future directions. *Nat Rev Drug Discov* 6: 75–92

16 Louis E, El Ghoul Z, Vermeire S, Dall'Ozzo S, Rutgeerts P, Paintaud G, Belaiche J, De Vos M, Van Gossum A, Colombel J et al (2005) Association between polymorphism in IgG Fc receptor IIIa coding gene and biological response to infliximab in Crohn's disease. *Inflamm Bowel Dis* 11: 75–76

17 Anolik J, Campbell D, Felgar R, Young F, Sanz I, Rosenblatt J, Looney R (2003) The relationship of FcgammaRIIIa genotype to degree of B cell depletion by rituximab in the treatment of systemic lupus erythematosus. *Arthritis Rheum* 48: 455–459

18 Tutuncu Z, Kavanaugh A, Zvaifler N, Corr M, Deutsch R, Boyle D (2005) Fc-gamma receptor type IIIA polymorphisms influence treatment outcomes in patients with inflammatory arthritis treated with tumor necrosis factor alpha-blocking agents. *Arthritis Rheum* 52: 2693–2696

19 Kastbom A, Bratt J, Ernestam S, Lampa J, Padyukov L, Soderkvist P, Skogh T (2007) Fcgamma receptor type IIIA genotype and response to tumor necrosis factor alpha-blocking agents in patients with rheumatoid arthritis. *Arthritis Rheum* 56: 448–452

20 Roll P, Palanichamy A, Kneitz C, Dorner T, Tony HP (2006) Regeneration of B cell subsets after transient B cell depletion using anti-CD20 antibodies in rheumatoid arthritis. *Arthritis Rheum* 54: 2377–2386

21 Silverman GJ (2006) Therapeutic B cell depletion and regeneration in rheumatoid arthritis: emerging patterns and paradigms. *Arthritis Rheum* 54: 2356–2367

22 Koshiba T, Kitade H, Van Damme B, Giulietti A, Overbergh L, Mathieu C, Waer M, Pirenne J (2003) Regulatory cell-mediated tolerance does not protect against chronic rejection. *Transplantation* 76: 588–596

23 Alexander JW, Light JA, Donaldson LA, Delmonico FL, Diethelm AG, Wilkinson A, Rosenthal JT, Thistlethwaite JR, Hunsicker LG, Matas AJ et al (1999) Evaluation of pre- and posttransplant donor-specific transfusion/cyclosporine A in non-HLA identical living donor kidney transplant recipients. Cooperative Clinical Trials in Transplantation Research Group. *Transplantation* 68: 1117–1124

24 Beyersdorf N, Gaupp S, Balbach K, Schmidt J, Toyka KV, Lin CH, Hanke T, Hunig T, Kerkau T, Gold R (2005) Selective targeting of regulatory T cells with CD28 superagonists allows effective therapy of experimental autoimmune encephalomyelitis. *J Exp Med* 202: 445–455

25 Mason D (1998) A very high level of crossreactivity is an essential feature of the T-cell receptor. *Immunol Today* 19: 395–404

26 Yang H, Welsh RM (1986) Induction of alloreactive cytotoxic T cells by acute virus infection of mice. *J Immunol* 136: 1186–1193

27 Braciale TJ, Andrew ME, Braciale VL (1981) Simultaneous expression of H-2-restricted and alloreactive recognition by a cloned line of influenza virus-specific cytotoxic T lymphocytes. *J Exp Med* 153: 1371–1376

28 Pasare C, Medzhitov R (2003) Toll pathway-dependent blockade of $CD4^+CD25^+$ T cell-mediated suppression by dendritic cells. *Science* 299: 1033–1036

29 Adams AB, Williams MA, Jones TR, Shirasugi N, Durham MM, Kaech SM, Wherry EJ,

Onami T, Lanier JG, Kokko KE et al (2003) Heterologous immunity provides a potent barrier to transplantation tolerance. *J Clin Invest* 111: 1887–1895

30 Lombardi G, Sidhu S, Daly M, Batchelor JR, Makgoba W, Lechler RI (1990) Are primary alloresponses truly primary? *Int Immunol* 2: 9–13

31 Lanzavecchia A, Sallusto F (2000) From synapses to immunological memory: the role of sustained T cell stimulation. *Curr Opin Immunol* 12: 92–98

32 London CA, Lodge MP, Abbas AK (2000) Functional responses and costimulator dependence of memory CD4⁺ T cells. *J Immunol* 164: 265–272

33 Wu Z, Bensinger SJ, Zhang J, Chen C, Yuan X, Huang X, Markmann JF, Kassaee A, Rosengard BR, Hancock WW et al (2004) Homeostatic proliferation is a barrier to transplantation tolerance. *Nat Med* 10: 87–92

34 Valujskikh A, Pantenburg B, Heeger PS (2002) Primed allospecific T cells prevent the effects of costimulatory blockade on prolonged cardiac allograft survival in mice. *Am J Transplant* 2: 501–509

35 Zhai Y, Meng L, Gao F, Busuttil RW, Kupiec-Weglinski JW (2002) Allograft rejection by primed/memory CD8⁺ T cells is CD154 blockade resistant: therapeutic implications for sensitized transplant recipients. *J Immunol* 169: 4667–4673

36 Hunig T, Dennehy K (2005) CD28 superagonists: mode of action and therapeutic potential. *Immunol Lett* 100: 21–28

37 Marshall E (2006) Drug trials. Violent reaction to monoclonal antibody therapy remains a mystery. *Science* 311: 1688–1689

38 Suntharalingam G, Perry MR, Ward S, Brett SJ, Castello-Cortes A, Brunner MD, Panoskaltsis N (2006) Cytokine storm in a phase 1 trial of the anti-CD28 monoclonal antibody TGN1412. *N Engl J Med* 355: 1018–1028

39 Fishwild D, Hudson D, Deshpande U, Kung A (1999) Differential effects of administration of a human anti-CD4 monoclonal antibody, HM6G, in nonhuman primates. *Clin Immunol* 92: 138–152

40 Greenwood J, Clark M, Waldmann H (1993) Structural motifs involved in human IgG antibody effector functions. *Eur J Immunol* 23: 1098–1104

41 Wu J, Edberg J, Redecha P, Bansal V, Guyre P, Coleman K, Salmon J, Kimberly R (1997) A novel polymorphism of FcgammaRIIIa (CD16) alters receptor function and predisposes to autoimmune disease. *J Clin Invest* 100: 1059–1070

42 Junghans RP, Anderson CL (1996) The protection receptor for IgG catabolism is the beta2-microglobulin-containing neonatal intestinal transport receptor. *Proc Natl Acad Sci USA* 93: 5512–5516

43 Isaacs J, Greenwood J, Waldmann H (1998) Therapy with monoclonal antibodies. II. The contribution of Fcg receptor binding and the influence of Ch1 and Ch3 domains on *in vivo* effector function. *J Immunol* 161: 3863–3869

44 Ghetie V, Popov S, Borvak J, Radu C, Matesoi D, Medesan C, Ober RJ, Ward E (1997) Increasing the serum persistence of an IgG fragment by random mutagenesis. *Nat Biotechnol* 15: 637–640

45 Shultz LD, Ishikawa F, Greiner DL (2007) Humanized mice in translational biomedical research. *Nat Rev Immunol* 7: 118–130
46 Isaacs J, Wing M, Greenwood J, Hazleman B, Hale G, Waldmann H (1996) A therapeutic human IgG4 monoclonal antibody that depletes target cells in humans. *Clin Exp Immunol* 106: 427–433
47 Wolbink GJ, Vis M, Lems W, Voskuyl AE, de Groot E, Nurmohamed MT, Stapel S, Tak PP, Aarden L, Dijkmans B (2006) Development of antiinfliximab antibodies and relationship to clinical response in patients with rheumatoid arthritis. *Arthritis Rheum* 54: 711–715
48 Maini RN, Breedveld FC, Kalden JR, Smolen JS, Davis D, Macfarlane JD, Antoni C, Leeb B, Elliott MJ, Woody JN et al (1998) Therapeutic efficacy of multiple intravenous infusions of anti-tumor necrosis factor alpha monoclonal antibody combined with low-dose weekly methotrexate in rheumatoid arthritis. *Arthritis Rheum* 41: 1552–1563
49 Strand V, Lipsky PE, Cannon GW, Calabrese LH, Wiesenhutter C, Cohen SB, Olsen NJ, Lee ML, Lorenz TJ, Nelson B (1993) Effects of administration of an anti-CD5 plus immunoconjugate in rheumatoid arthritis. Results of two Phase II studies. The CD5 Plus Rheumatoid Arthritis Investigators Group. *Arthritis Rheum* 36: 620–630
50 Benjamin R, Cobbold S, Clark M, Waldmann H (1986) Tolerance to rat monoclonal antibodies. Implications for serotherapy. *J Exp Med* 163: 1539–1552
51 Gilliland L, Walsh L, Frewin M, Wise M, Tone M, Hale G, Kioussis D, Waldmann H (1999) Elimination of the immunogenicity of therapeutic antibodies. *J Immunol* 162: 3663–3671
52 Weinblatt ME, Keystone EC, Furst DE, Moreland LW, Weisman MH, Birbara CA, Teoh LA, Fischkoff SA, Chartash EK (2003) Adalimumab, a fully human anti-tumor necrosis factor alpha monoclonal antibody, for the treatment of rheumatoid arthritis in patients taking concomitant methotrexate: the ARMADA trial. *Arthritis Rheum* 48: 35–45
53 Hanauer S, Feagan B, Lichtenstein G, Mayer L, Schreiber S, Colombel J, Rachmilewitz D, Wolf D, Olson A, Bao W et al (2002) Maintenance infliximab for Crohn's disease: the ACCENT I randomised trial. *Lancet* 359: 1541–1549
54 Rutgeerts P, D'Haens G, Targan S, Vasiliauskas E, Hanauer S, Present D, Mayer L, Van Hogezand R, Braakman T, DeWoody K et al (1999) Efficacy and safety of retreatment with anti-tumor necrosis factor antibody (infliximab) to maintain remission in Crohn's disease. *Gastroenterology* 117: 761–769
55 Clark M (2000) Antibody humanization: a case of the "Emperor's new clothes"? *Immunol Today* 21: 397–402
56 Gao W, Lu Y, El Essawy B, Oukka M, Kuchroo VK, Strom TB (2007) Contrasting effects of cyclosporine and rapamycin in *de novo* generation of alloantigen-specific regulatory T cells. *Am J Transplant* 7: 1722–1732
57 Jenkins MK, Chen CA, Jung G, Mueller DL, Schwartz RH (1990) Inhibition of antigen-specific proliferation of type 1 murine T cell clones after stimulation with immobilized anti-CD3 monoclonal antibody. *J Immunol* 144: 16–22
58 Woodside KJ, Hu M, Liu Y, Song W, Hunter GC, Daller JA (2006) Apoptosis of allo-

specifically activated human helper T cells is blocked by calcineurin inhibition. *Transpl Immunol* 15: 229–234

59 Qu Y, Zhang B, Zhao L, Liu G, Ma H, Rao E, Zeng C, Zhao Y (2007) The effect of immunosuppressive drug rapamycin on regulatory $CD4^+CD25^+Foxp3^+$ T cells in mice. *Transpl Immunol* 17: 153–161

60 Goekoop-Ruiterman YP, de Vries-Bouwstra JK, Allaart CF, van Zeben D, Kerstens PJ, Hazes JM, Zwinderman AH, Ronday HK, Han KH, Westedt ML et al (2005) Clinical and radiographic outcomes of four different treatment strategies in patients with early rheumatoid arthritis (the BeSt study): a randomized, controlled trial. *Arthritis Rheum* 52: 3381–3390

61 Chatenoud L (2006) Immune therapies of autoimmune diseases: are we approaching a real cure? *Curr Opin Immunol* 18: 710–717

62 Network It www.immunetolerance.org

63 Devlin J, Doherty D, Thomson L, Wong T, Donaldson P, Portmann B, Williams R (1998) Defining the outcome of immunosuppression withdrawal after liver transplantation. *Hepatology* 27: 926–933

64 Mazariegos GV, Reyes J, Marino IR, Demetris AJ, Flynn B, Irish W, McMichael J, Fung JJ, Starzl TE (1997) Weaning of immunosuppression in liver transplant recipients. *Transplantation* 63: 243–249

65 Zoller KM, Cho SI, Cohen JJ, Harrington JT (1980) Cessation of immunosuppressive therapy after successful transplantation: a national survey. *Kidney Int* 18: 110–114

66 Roussey-Kesler G, Giral M, Moreau A, Subra JF, Legendre C, Noel C, Pillebout E, Brouard S, Soulillou JP (2006) Clinical operational tolerance after kidney transplantation. *Am J Transplant* 6: 736–746

67 Calne R, Friend P, Moffatt S, Bradley A, Hale G, Firth J, Bradley J, Smith K, Waldmann H (1998) Prope tolerance, perioperative campath 1H, and low-dose cyclosporin monotherapy in renal allograft recipients. *Lancet* 351: 1701–1702

Index

AChR (nicotinic acetylcholine receptor) 40
activation-induced cell death (AICD) 160
acute lymphoblastic leukemia 113
acute myeloid leukemia1 (AML1) 149
airway response 114
Ala-Ala 60
alloantibody 111
allograft rejection 41, 43
alloreactive CD8 T cell 74, 76, 77
alloreactive memory cell 76
alloreactivity 41
altered peptide ligand (APL) 16, 20, 35, 37
AML1 (acute myeloid leukemia1) 149
anergy 21, 36, 129, 131, 162
antagonism 35, 37
antigen-presenting cell (APC) 1
APL 16, 20, 35, 37
APL, dominant negative signaling cascade, APL 37
asthma 110, 117
autoimmune diabetes 65
autoimmune disease 17, 61
autoimmunity 38, 43

B cell 111, 116
B7 family 93
beta-cell regeneration 65
bimosiamose 117
bimosiamose (sLe^x mimetic) 117
biological therapy registry 179
bone marrow transplantation 110

$[Ca^{2+}]_i$ 157
Ca^{2+} response 1
Ca^{2+}-calcineurin 160
calcineurin inhibition 113
Casitas B-lineage lymphoma (Cbl-b) 132, 135, 137, 158
CBLB 158
Cbl-b, functions 137
Cbl-b, regulation 137
Cbl-b, structure 135
CCR7 4
CD2 16, 107
CD2, anti-CD2 monoclonal antibody 107, 113
CD2, anti-CD2 therapy 107
CD3 16
CD3ζ 38
CD3, aglycosylated humanized anti-CD3 IgG_1 (campath 3 or ChAglyCD3) 60
CD3, anti-CD3 expanded Tregs 63, 65
CD3, anti-CD3 mAbs, effect on EAE 64
CD3, anti-CD3 monoclonal antibody 58
CD-3 specific antibody 58
CD4 16
$CD4^+$ T cell 159
$CD4^+CD25^-$ 149
$CD4^+CD25^+$ Treg cell 78, 148, 160
CD4, anti-CD4 monoclonal antibody 20
$CD4^+$ Tregs, expansion of adaptive 64

CD8+ T cell 73, 147, 159
CD11a (LFA-1) 109, 110
CD18/CD11a 110
CD22 107
CD25 155
CD28 16, 112, 129
CD40 71, 73
CD40L (CD154) 16, 71, 73, 75, 77, 112
CD40L, anti-CD40L monoclonal antibody 113
CD40L, endothelial activation 73
CD40L blockade 71, 74-77
CD45 16
CD48 107
CD80 162
CD83 107
CD103 21
CDP850 (humanized mAb) 117
central nervous system 115
CHL-1 107
chromatin-remodeling factor 150
chronic berylliosis 114
chronic rejection 41
clinical trial 176–178
clinical trial, design 176, 177
clinical trial, evaluation 177, 178
copolymer-I (COP-I) 148
co-receptor blockade 51, 52, 54
co-stimulation blockade 54
CREB/CREM 132
Ctla4 155
CTLA-4 40, 88
CY1503 117
CY1747 (mAb) 117
cyclosporin A (CsA) 57, 110
cytohesin-1 112

DAGKα (diacylglycerol kinase α) 132
Dby 42
dendritic cell (DC) 1, 18, 22, 88, 162
determinant spreading, autoimmune response 40, 41
diacylglycerol kinase α (DAGKα) 132
diapedesis 111
1α,25-dihydroxyvitamin D3 (vitD3) 19, 20
dominant tolerance 15
donor chimerism 51
DREG200 117

E1 or ubiquitin-activating enzyme 133
E2 or ubiquitin-conjugating enzyme 133
E3 or ubiquitin-protein ligase 133, 135
EAE (experimental autoimmune encephalomyelitis) 39, 64, 161
early growth response protein (Egr) 137
efalizumab 109, 110, 113, 114
efomycine (sLe^x mimetic) 117
enlimomab 108
eosinophil 114, 115
ERK 38, 40
Erk1/2 112
everolimus 113
exendin-4 65
experimental autoimmune encephalomyelitis (EAE) 39, 64, 161

flu-like syndrome 58, 64
forkhead 149
FOXP1 149
FOXP2 149
FOXP3 19, 21, 40, 147, 149, 150
FOXP3, post-translational modification of FOXP3 150
Foxp3 155
FOXP4 149

Gata3 157
gene related to anergy in lymphocytes (GRAIL) 132, 139, 140
glomerulonephritis 117
glucocorticosteroid 114
graft *versus* host disease 42
GRAIL (gene related to anergy in lymphocytes) 132, 139, 140
GRAIL, function of 140

GRAIL, regulation of 140
GRAIL, structure of 139
GW559090X $\alpha_4\beta_1$ (VLA-4) antagonist 110

HA-1 41, 42
HDAC7 150
HDAC (histone deacetylase) 150
HDAC9 150
heart transplantation 109
HECT 133
heme-oxygenase-1 19
hepatocellular transplantation 109
heteromerization 149
histone acetyltransferase 150
histone deacetylase (HDAC) 150
histone modification enzyme 150
homoassociation 149
homotetramerization 149
HuEP5C7 (humanized mAb) 117
human CD3 (T3/4.A) 61
humanized anti-LFA-1 monoclonal antibody (efalizumab) 109, 110, 113, 114
humanized IgG_1 form of OKT3 [hOkt3γ1 (Ala-Ala)] 60
humanized IgG_2 anti-CD3 (HuM291 or visilizumab) 61
humanized neutralizing IgG4κ mAbs 115

ICAM, anti-ICAM therapy 107
ICAM, anti-ICAM-1 monoclonal antibody 113
ICAM-1 16, 108, 109, 113
ICAM-2 108, 109
ICAM-3 109
IDO (indoleamine 2,3-dioxygenase) 19, 89, 162
Ifng 157
IFN-γ 89, 159, 148
Ikaros 132
IL-6 148
IL-10 19, 21
immunogenicity 175, 176
immune deviation 36, 38, 39
immune intervention 57
immune suppression 41
immune tolerance, biomarkers of 178
immune tolerance, naturally occurring 179
immune tolerance, spontaneous 179
immune tolerance, prope 180
immunoglobulin supergene family (IgSF) cell adhesion molecule (CAM) 107
indoleamine 2,3-dioxygenase (IDO) 19, 89, 162
infectious tolerance 16, 51, 52
inflammatory bowel disease 115
integrin 107, 108, 110, 111, 115
α_4 integrin 108, 110, 115
$\alpha_4\beta_7$ integrin
$\alpha_4\beta_7$ integrin (VLA-4) 108
$\alpha_L\beta_2$ (CD11a/CD18 or LFA-1) integrin 107
$\alpha_M\beta_2$ (CD11b/CD18 or Mac-1) integrin 107
β_2 integrin family 107
intracellular adhesion molecule (ICAM) 16, 107–109, 112–114
I-R following cardiopulmonary bypass 117
I-R following pulmonary thromboendarte-tectomy 117
I-R injury 117
ischemia reperfusion injury (I-R injury) 73, 117
islet transplantation 109
Itch 132, 138
Itch, function of 139
Itch, regulation of 138
Itch, structure of 138

JAM-1 109
JC virus-induced progressive multifocal leukoencephalopathy 115
JNK 38
junctional adhesion molecule (JAM) 108, 109

kidney allograft rejection 113
kidney transplantation 113

kinetic discrimination model 37
kynurenine 92

L1-CAM 107
leukocyte rolling 116
LFA-1 16, 108, 109, 111–114, 159
LFA-1, anti-LFA-1 monoclonal antibody 112–114
LFA-1, anti-LFA-1 therapy 112
LFA-1-deficient mouse 112
LFA-1-ICAM-1 interaction 112–114
linked suppression 16, 18, 19, 22, 39, 40, 42, 43, 51, 52
lymph node (LN) 3
lymphocyte motility 2
lymphocyte trafficking 112

MAd-CAM-1 (mucosal adressin cell adhesion molecule-1) 108
MBP (myelin basic protein) 38
memory T cell 111
minor histocompatibility (mH) antigen 41
mixed hematopoietic chimerism 19
MOG (myelin oligodendrocyte glycoprotein) 41
mucosal adressin cell adhesion molecule-1 (MAd-CAM-1) 108
multiple sclerosis 38, 39, 109, 110, 115
myasthenia gravis (MG) 40
myelin basic protein (MBP) 38
myelin oligodendrocyte glycoprotein (MOG) 41
myocardial infarction 110, 117
myocardial injury 117
myocardial ischemia-reperfusion 108

naïve T cell 111
natalizumab 115
natalizumab (targets $\alpha_4\beta_7$, $\alpha_4\beta_1$) 110
natural killer (NK) cell 116
NCAM (neural cell adhesion molecule) 107
neural cell adhesion molecule (NCAM) 107
neutrophil 112
NFAT (nuclear factor of activated T cells) 131, 149
NFATc2 149
nicotinic acetylcholine receptor (AChR) 40
NOD mouse 62
non-Fc binding anti-CD3 60
nuclear factor of activated T cells (NFAT) 131, 149

OC229648 (sLe^x mimetic) 117
OKT3 59
oral tolerance 21
Orthoklone 59
otubain1 140

p27^{kip1} 132
p38 38
partial agonist 36, 38-40, 42
PECAM-1 (platelet-endothelial cell adhesion molecule) 107, 108
peritonitis 117
phosphatidylinositol 3-kinase (PI3K) 137
PI3K (phosphatidylinositol 3-kinase) 137
PKC 162
PKC-θ 139
platelet-endothelial cell adhesion molecule (PECAM-1) 107, 108
PLC-γ1 139
post-transplant lymphoproliferative disease 110, 114
preclinical study 174
preclinical testing 171–174
preclinical testing, animal models, extrinsic limitations 173, 174
preclinical testing, animal models, intrinsic limitations 172, 173
privileged environment, graft 52
pro-inflammatory cytokine 72
pro-inflammatory response 73
proinsulin vaccine, in combination with anti-CD3 145-2C11 mAb 65

proteasome 134
proteolipid protein 39
psoriasis 110, 114, 117
psoriatic arthritis 60

Qa-1 147

RAGE (receptor for advance glycation end products) 108
rat insulin promoter-lymphocyte choriomeningitis virus (RIP-LCMV) mouse, transgenic 62
receptor for advance glycation end products (RAGE) 108
regulation, T cell-mediated 50, 51
regulatory T cell (Treg cell) 8, 15, 19–22, 39, 42, 43, 52, 78, 88, 147, 155
renal transplantation 58, 108, 110, 117
research network 179
resistance 49, 51, 52
rheumatoid arthritis 108
Rho guanine dissociation inhibitor (RhoGDI) 140
RING 133
rolling lymphocyte 111
RORa 150
RORγt 157
rovelizumab 110
rPSGL-Ig 117
Runx1 149

selectin 115–117
P-, E- and L-selectin 116, 117
selectin inhibitor 116
short-course treatment 64
sialic acid binding Ig-like lectin (SIGLEC) 107
SIGLEC (sialic acid binding Ig-like lectin) 107
sLe^x 116
SMAD3 158
SMAD4 160
stroke 108, 117
suppressor T cell 147

T-bet 157
T cell homeostasis 155
T cell receptor (TCR) 35, 111, 112, 129
tetramerization 149
Tgfbr2 160
TGF-β 20, 148
TGF-β1 40, 155
Th1 115, 157
Th1-type immune response 72
Th2 115, 157
Th2 cytokine 113
Th3 cell 158
Th17 cell 157
therapeutic target, identification of 171
thymic epithelial cell (TEC) 22
TIP60 150
tissue injury 73
TLR (toll-like receptor) 76
TNF 148
TNFRII, anti-TNFRII monoclonal antibody 148
tolerance 49-54, 62, 155, 178–180
tolerance, transplantation 51, 52
tolerance induction 57
tolerance mediated by regulatory T cells (Tregs) 62
tolerogen 49
toll-like receptor (TLR) 76
Tr1 21
tracheal transplantation 109
transforming growth factor (TGF)-β1 40, 155
Treg cell 8, 15, 19–22, 39, 42, 43, 52, 78, 88, 147, 155
Treg cell, modulation of function 151
Treg cell conversion 20
Treg specificity 22
Treg cell marker 52
tryptophan catabolism 91
two-photon microscopy 3
type 1 diabetes 61, 160
type 1 diabetes and inflammatory bowel disease 160

ubiquitin 133

vaccine development 151
vascular cell adhesion molecule
(VCAM-1) 107, 108
vascular endothelial cell 73
vascular endothelium 111
Vav1 137
VCAM-1 (vascular cell adhesion molecule) 107, 108
visilizumab 61
vitD31 (α,25-dihydroxyvitamin D3) 19, 20
VLA-4 108, 115

xenoantibody 111
X-linked autoimmunity-allergic dysregulation syndrome (XLAAD) 148

ZAP-70 38

The PIR-Series
Progress in Inflammation Research

Homepage: http://www.birkhauser.ch

Up-to-date information on the latest developments in the pathology, mechanisms and therapy of inflammatory disease are provided in this monograph series. Areas covered include vascular responses, skin inflammation, pain, neuroinflammation, arthritis cartilage and bone, airways inflammation and asthma, allergy, cytokines and inflammatory mediators, cell signalling, and recent advances in drug therapy. Each volume is edited by acknowledged experts providing succinct overviews on specific topics intended to inform and explain. The series is of interest to academic and industrial biomedical researchers, drug development personnel and rheumatologists, allergists, pathologists, dermatologists and other clinicians requiring regular scientific updates.

Available volumes:
T Cells in Arthritis, P. Miossec, W. van den Berg, G. Firestein (Editors), 1998
Chemokines and Skin, E. Kownatzki, J. Norgauer (Editors), 1998
Medicinal Fatty Acids, J. Kremer (Editor), 1998
Inducible Enzymes in the Inflammatory Response,
D.A. Willoughby, A. Tomlinson (Editors), 1999
Cytokines in Severe Sepsis and Septic Shock, H. Redl, G. Schlag (Editors), 1999
Fatty Acids and Inflammatory Skin Diseases, J.-M. Schröder (Editor), 1999
Immunomodulatory Agents from Plants, H. Wagner (Editor), 1999
Cytokines and Pain, L. Watkins, S. Maier (Editors), 1999
In Vivo *Models of Inflammation*, D. Morgan, L. Marshall (Editors), 1999
Pain and Neurogenic Inflammation, S.D. Brain, P. Moore (Editors), 1999
Anti-Inflammatory Drugs in Asthma, A.P. Sampson, M.K. Church (Editors), 1999
Novel Inhibitors of Leukotrienes, G. Folco, B. Samuelsson, R.C. Murphy (Editors), 1999
Vascular Adhesion Molecules and Inflammation, J.D. Pearson (Editor), 1999
Metalloproteinases as Targets for Anti-Inflammatory Drugs,
K.M.K. Bottomley, D. Bradshaw, J.S. Nixon (Editors), 1999
Free Radicals and Inflammation, P.G. Winyard, D.R. Blake, C.H. Evans (Editors), 1999
Gene Therapy in Inflammatory Diseases, C.H. Evans, P. Robbins (Editors), 2000
New Cytokines as Potential Drugs, S. K. Narula, R. Coffmann (Editors), 2000
High Throughput Screening for Novel Anti-inflammatories, M. Kahn (Editor), 2000
Immunology and Drug Therapy of Atopic Skin Diseases,
C.A.F. Bruijnzeel-Komen, E.F. Knol (Editors), 2000
Novel Cytokine Inhibitors, G.A. Higgs, B. Henderson (Editors), 2000
Inflammatory Processes. Molecular Mechanisms and Therapeutic Opportunities,
L.G. Letts, D.W. Morgan (Editors), 2000
Cellular Mechanisms in Airways Inflammation, C. Page, K. Banner, D. Spina (Editors), 2000
Inflammatory and Infectious Basis of Atherosclerosis, J.L. Mehta (Editor), 2001
Muscarinic Receptors in Airways Diseases, J. Zaagsma, H. Meurs, A.F. Roffel (Editors), 2001

TGF-β and Related Cytokines in Inflammation, S.N. Breit, S. Wahl (Editors), 2001
Nitric Oxide and Inflammation, D. Salvemini, T.R. Billiar, Y. Vodovotz (Editors), 2001
Neuroinflammatory Mechanisms in Alzheimer's Disease. Basic and Clinical Research, J. Rogers (Editor), 2001
Disease-modifying Therapy in Vasculitides, C.G.M. Kallenberg, J.W. Cohen Tervaert (Editors), 2001
Inflammation and Stroke, G.Z. Feuerstein (Editor), 2001
NMDA Antagonists as Potential Analgesic Drugs, D.J.S. Sirinathsinghji, R.G. Hill (Editors), 2002
Migraine: A Neuroinflammatory Disease? E.L.H. Spierings, M. Sanchez del Rio (Editors), 2002
Mechanisms and Mediators of Neuropathic pain, A.B. Malmberg, S.R. Chaplan (Editors), 2002
Bone Morphogenetic Proteins. From Laboratory to Clinical Practice, S. Vukicevic, K.T. Sampath (Editors), 2002
The Hereditary Basis of Allergic Diseases, J. Holloway, S. Holgate (Editors), 2002
Inflammation and Cardiac Diseases, G.Z. Feuerstein, P. Libby, D.L. Mann (Editors), 2003
Mind over Matter – Regulation of Peripheral Inflammation by the CNS, M. Schäfer, C. Stein (Editors), 2003
Heat Shock Proteins and Inflammation, W. van Eden (Editor), 2003
Pharmacotherapy of Gastrointestinal Inflammation, A. Guglietta (Editor), 2004
Arachidonate Remodeling and Inflammation, A.N. Fonteh, R.L. Wykle (Editors), 2004
Recent Advances in Pathophysiology of COPD, P.J. Barnes, T.T. Hansel (Editors), 2004
Cytokines and Joint Injury, W.B. van den Berg, P. Miossec (Editors), 2004
Cancer and Inflammation, D.W. Morgan, U. Forssmann, M.T. Nakada (Editors), 2004
Bone Morphogenetic Proteins: Bone Regeneration and Beyond, S. Vukicevic, K.T. Sampath (Editors), 2004
Antibiotics as Anti-Inflammatory and Immunomodulatory Agents, B.K. Rubin, J. Tamaoki (Editors), 2005
Antirheumatic Therapy: Actions and Outcomes, R.O. Day, D.E. Furst, P.L.C.M. van Riel, B. Bresnihan (Editors), 2005
Regulatory T-Cells in Inflammation, L. Taams, A.N. Akbar, M.H.M Wauben (Editors), 2005
Sodium Channels, Pain, and Analgesia, K. Coward, M. Baker (Editors), 2005
Turning up the Heat on Pain: TRPV1 Receptors in Pain and Inflammation, A.B Malmberg, K.R. Bley (Editors), 2005
The NPY Family of Peptides in Immune Disorders, Inflammation, Angiogenesis and Cancer, Z. Zukowska, G. Z. Feuerstein (Editors), 2005
Toll-like Receptors in Inflammation, L.A.J. O'Neill, E. Brint (Editors), 2005
Complement and Kidney Disease, P. F. Zipfel (Editor), 2006
Chemokine Biology – Basic Research and Clinical Application, Volume 1: Immunobiology of Chemokines, B. Moser, G. L. Letts, K. Neote (Editors), 2006
The Hereditary Basis of Rheumatic Diseases, R. Holmdahl (Editor), 2006
Lymphocyte Trafficking in Health and Disease, R. Badolato, S. Sozzani (Editors), 2006
In Vivo *Models of Inflammation, 2nd Edition, Volume I*, C.S. Stevenson, L.A. Marshall, D.W. Morgan (Editors), 2006
In Vivo *Models of Inflammation, 2nd Edition, Volume II*, C.S. Stevenson, L.A. Marshall, D.W. Morgan (Editors), 2006
Chemokine Biology – Basic Research and Clinical Application. Volume II: Pathophysiology of Chemikines, K. Neote, G.L. Letts, B. Moser (Editors), 2007
Adhesion Molecules: Function and Inhibition, K. Ley (Editor), 2007

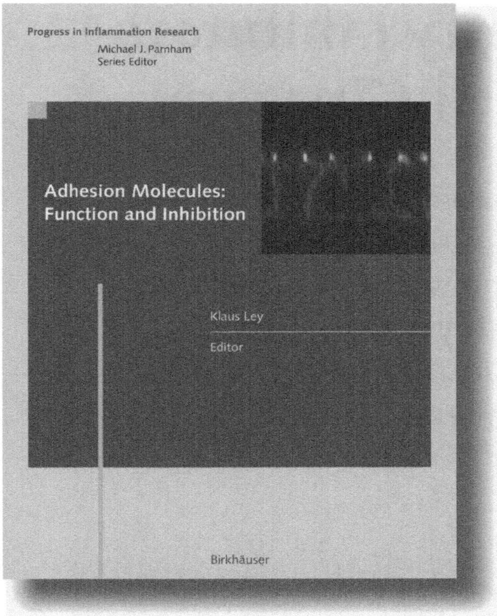

Adhesion Molecules: Function and Inhibition

Klaus Ley (ed.)
La Jolla Institute for Allergy & Immunology, La Jolla, CA, USA

2007. XIV, 309 pp. 23 illus. Hardcover
ISBN 978-3-7643-7974-2
PIR – Progress in Inflammation Research

Inflammatory cell recruitment requires the concerted action of at least five major sets of adhesion molecules: integrins, immunoglobulin-like molecules, selectins, carbohydrate structures serving as selectin ligands, and certain ectoenzymes. This volume gives a comprehensive overview on the most relevant leukocyte and endothelial adhesion molecules.
The chapters are written by leaders in the field and focus on the biology, structure, function, and regulation of adhesion molecules. Currently approved adhesion molecule-based therapies are reviewed and an outlook for future approaches is also provided.
The book is of interest to clinicians and scientists from immunology, physiology, cancer research, rheumatology, allergology, infectious diseases, gastroenterology, pulmonology and cardiology.

From the contents:
P-selectin glycoprotein ligand-1 (PSGL-1).- L-selectin-mediated leukocyte adhesion and migration.- Endothelial ICAM-1 functions in adhesion and signaling during leukocyte recruitment.- alpha4-integrins: structure, function and secrets.- VCAM-1 and its functions in development and inflammatory diseases.- Lymphocyte function-associated antigen-1 (LFA-1) and macrophage antigen-1 (Mac-1): Cooperative partners in leukocyte emigration and function.- PECAM: Regulating the start of diapedesis.- Role of alpha6beta1 integrin in leukocyte adhesion and transmigration.- Vascular adhesion protein-1 (VAP-1).- The role of ESAM in neutrophil emigration into inflamed tissues.- Structure and function of JAM proteins.- Promises and limitations of targeting adhesion molecules for therapy

Birkhäuser
Viaduktstrasse 42
CH-4051 Basel

Tel. +41 61 205 07 77
e-mail: sales@birkhauser.ch
www.birkhauser.ch

GPSR Compliance

The European Union's (EU) General Product Safety Regulation (GPSR) is a set of rules that requires consumer products to be safe and our obligations to ensure this.

If you have any concerns about our products, you can contact us on ProductSafety@springernature.com

In case Publisher is established outside the EU, the EU authorized representative is:

Springer Nature Customer Service Center GmbH
Europaplatz 3
69115 Heidelberg, Germany

Batch number: 10578049

Printed by Printforce, the Netherlands